大数据及人工智能产教融合系列丛书

大数据专业英语

高文宇　徐　亭　主编

电子工业出版社
Publishing House of Electronics Industry
北京·BEIJING

内 容 简 介

本书介绍了大数据的概念、特点、市场、技术、分析、应用，还专门讨论了云计算、人工智能和区块链等热点新技术及应用，教学设计层次清晰，每个单元都遵循同样的编排体系，内容图文并茂，对口语技能、阅读技能、翻译技能的学习任务进行了合理的设计。而且，本书还对每个单元的教学重点和专业词汇进行了注释，设置了基于内容的阅读理解练习、词汇练习和翻译练习。

本书是为高等院校相关专业（尤其是数据科学与大数据技术、大数据管理与应用、信息管理、经济管理等专业开设的大数据相关课程）设计并编写的具有丰富实践特色的教材。本书也可作为有一定实践经验的 IT 应用人员、管理人员的参考书，以及继续教育的教材和英语类专业科技英语课程的选修教材。

未经许可，不得以任何方式复制或抄袭本书之部分或全部内容。
版权所有，侵权必究。

图书在版编目（CIP）数据

大数据专业英语 / 高文宇，徐亭主编. —北京：电子工业出版社，2020.10
（大数据及人工智能产教融合系列丛书）
ISBN 978-7-121-39629-8

Ⅰ. ①大… Ⅱ. ①高… ②徐… Ⅲ. ①数据处理－英语－高等学校－教材 Ⅳ. ①TP274

中国版本图书馆 CIP 数据核字（2020）第 179280 号

责任编辑：李　冰
文字编辑：冯　琦
印　　刷：涿州市般润文化传播有限公司
装　　订：涿州市般润文化传播有限公司
出版发行：电子工业出版社
　　　　　北京市海淀区万寿路 173 信箱　　邮编：100036
开　　本：787×1 092　1/16　印张：12.75　字数：340 千字
版　　次：2020 年 10 月第 1 版
印　　次：2024 年 7 月第 2 次印刷
定　　价：68.00 元

凡所购买电子工业出版社图书有缺损问题，请向购买书店调换。若书店售缺，请与本社发行部联系，联系及邮购电话：（010）88254888，88258888。
质量投诉请发邮件至 zlts@phei.com.cn，盗版侵权举报请发邮件至 dbqq@phei.com.cn。
本书咨询联系方式：fengq@phei.com.cn。

编 委 会

（按姓氏音序排列）

总顾问

郭华东　中国科学院院士
谭建荣　中国工程院院士

编委会主任

韩亦舜

编委会副主任

孙　雪　徐　亭　赵　强

编委会成员

薄智泉　卜　辉　陈晶磊　陈　军　陈新刚　杜晓梦
高文宇　郭　炜　黄代恒　黄枝铜　李春光　李雨航
刘川意　刘　猛　单　单　盛国军　田春华　王薇薇
文　杰　吴垌沅　吴　建　杨　扬　曾　光　张鸿翔
张文升　张粤磊　周明星

《大数据专业英语》编委会

(按姓氏音序排列)

高级顾问

薄智泉　　戴炜栋　　高志凯　　郭毅可　　李　颉
卢建新　　明特·戴尔　容淳铭　　徐　来　　赵国栋
张延川　　章玉贵

编委会主任

高文宇　　徐　亭

编委会副主任

吴启锋　　郑运帧　　周冬祥

编委会成员

胡　立　　金圆圆　　梁诗悦　　陆云峰　　吕百花
沈一涵　　王丁桃　　王苗苗　　张世光　　张　欣

丛书推荐序一

数字经济的思维观与人才观

大数据的出现,给我们带来了巨大的想象空间:对科学研究来说,大数据已成为继实验、理论和计算模式之后的数据密集型科学范式的典型代表,带来了科研方法论的变革,正在成为科学发现的新引擎;对产业来说,在当今互联网、云计算、人工智能、大数据、区块链这些蓬勃发展的科技中,主角是数据,数据作为新的生产资料,正在驱动整个产业进行数字化转型。正因如此,大数据已成为知识经济时代的战略高地,数据主权已经成了继边防、海防、空防之后,另一个大国博弈的空间。

实现这些想象空间,需要构建众多大数据领域的基础设施,小到科学大数据方面的国家重大基础设施,大到跨越国界的"数字丝路""数字地球"。今天,我们看到大数据基础设施研究中心已经把人才纳入基础设施的范围,组织编写了这套丛书,这个视角是有意义的。新兴的产业需要相应的人才培养体系与之相配合,人才培养体系的建立往往存在滞后性。因此,尽可能缩窄产业人才需求和培养过程间的"缓冲带",将教育链、人才链、产业链、创新链衔接好,就是"产教融合"理念提出的出发点和落脚点。可以说,大数据基础设施研究中心为我国大数据、人工智能事业发展模式的实践迈出了较为坚实的一步,这个模式意味着数字经济宏观的可行路径。

作为我国首套大数据及人工智能方面的产教融合丛书,其以数据为基础,内容涵盖了数据认知与思维、数据行业应用、数据技术生态等各个层面及其细分方向,是数十个代表了行业前沿和实践的产业团队的知识沉淀。特别是在作者遴选时,这套丛书注重选择兼具产业界和学术界背景的行业专家,以便让丛书成为中国大数据知识的一次汇总,这对于中国数据思维的传播、数据人才的培养来说,是一个全新的范本。

我也期待未来有更多产业界的专家及团队加入本套丛书体系中,并和这套丛书共同更新迭代,共同传播数据思维与知识,夯实我国的数据人才基础设施。

郭华东
中国科学院院士

丛书推荐序二

产教融合打造创新人才培养的新模式

数字技术、数字产品和数字经济，是信息时代发展的前沿领域，不断迭代着数字时代的定义。数据是核心战略性资源，自然科学、工程技术和社科人文拥抱数据的力度，对于学科新的发展具有重要意义。同时，数字经济是数据的经济，既是各项高新技术发展的动力，又为传统产业转型提供了新的数据生产要素与数据生产力。

这套丛书从产教融合的角度出发，在整体架构上，涵盖了数据思维方式拓展、大数据技术认知、大数据技术高级应用、数据化应用场景、大数据行业应用、数据运维、数据创新体系七个方面，编写宗旨是搭建大数据的知识体系，传授大数据的专业技能，描述产业和教育相互促进过程中所面临的问题，并在一定程度上提供相应阶段的解决方案。丛书的内容规划、技术选型和教培转化由新型科研机构——大数据基础设施研究中心牵头，而场景设计、案例提供和生产实践由一线企业专家与团队贡献，两者紧密合作，提供了一个可借鉴的尝试。

大数据领域人才培养的一个重要方面，就是以产业实践为导向，以传播和教育为出口，最终服务于大数据产业与数字经济，为未来的行业人才树立技术观、行业观、产业观，进而助力产业发展。

这套丛书适用于大数据技能型人才的培养，适合作为高校、职业学校、社会培训机构从事大数据教学和研究的教材或参考书，对于从事大数据管理和应用的人员、企业信息化技术人员也有重要的参考价值。让我们一起努力，共同推进大数据技术的教学、普及和应用！

谭建荣
中国工程院院士
浙江大学教授

推荐序一

人类在传感器技术和计算机计算能力方面的进步催生了本书的主角——大数据（Big Data）！

大数据是继物联网、云计算后，IT产业的又一次颠覆性技术革命。物联网能实现快速精准的信息识别、管理和控制，云计算能实现资源共享和网络协同，而大数据是人类测量、记录和分析世界的直接手段。

近年来，随着物联网、云计算、移动互联网、车联网等技术的成熟和迅速普及，人类社会正在以更快的速度产生图像、音频、视频、健康档案等不同类别的可供采集和分析的海量数据。国际数据公司IDC预测，2025年全球数据量将达到175ZB。信息技术的发展进入"大智移云"时代，"大智移云"即大数据、智能化（包括物联网与人工智能）、移动互联网和云计算。"大智移云"技术将信息技术、通信技术与智能制造技术融合，标志着互联网从消费互联网向产业互联网拓展，深刻影响经济社会的方方面面。我们手握大数据之剑，就有了征服未来的本钱。

"大数据专业英语"是数据智能时代人才培养的核心课程之一。高文宇先生和徐亨先生联合主编的《大数据专业英语》的出版恰逢其时。读完书稿，我感觉本书颇有特点。首先，本书主要聚焦大数据的分析、应用，涉及云计算、医疗健康、金融、教育、零售、运输等，针对大众当下关注的领域，内容比较丰富；其次，本书不仅介绍了大数据技术，还专门讨论了人工智能和区块链等热点技术和应用；再次，本书在写作方法上结合对话和阅读，由浅入深、通俗易懂，可连贯阅读，也可有选择地关注其中的某些内容；最后，所有章节重点突出、练习丰富、图文并茂，非常方便自学和教学。

本书内容丰富且视角多样，虽然难以涉及所有方面，但不失为对写作方式的一种探索。希望本书的出版能够起到抛砖引玉的作用，帮助更多的人了解和掌握大数据技术，催生更多的大数据应用创新成果，促进我国数字经济的发展。总之，在现有众多与大数据英语有关的读物中，本书在内容和形式方面都很突出，值得一读。

李颉

上海交通大学教授

日本工程院外籍院士

IEEE 大数据委员会共同主席

推荐序二

"天地者，万物之逆旅；光阴者，百代之过客。"当前科技发展迅猛，珍惜时间、增强学习十分必要。应把创新摆在国家发展全局的核心位置，将推进经济数字化作为实现创新发展的重要动能。当前，以互联网、大数据、区块链、人工智能为代表的新一代信息技术日新月异，给各国经济社会发展、国家管理、社会治理、人民生活带来重大而深远的影响。中国要高度重视大数据发展，要通过大数据进行产业创新、打造数字经济、提升国家治理水平、改善民生及保障国家数据安全。

面对新时代的新要求，英语教育也踏上了新征程，不断改革和发展。应当着重培养适应新时代中国特色社会主义建设的各类英语人才，即具有丰富的专业知识、跨文化沟通能力、中国情怀、全球视野，能在"一带一路"建设、推动"中国文化走出去"、构建"人类命运共同体"中发挥重要作用的专业人才。英语学科的建设与发展尤其需要把握一流学科建设的契机，在师资建设、人才培养、科学研究、社会服务等方面开拓创新，实现可持续发展。大数据专业英语重在搜集、挖掘和整理，并找出关系、重点、规律，洞察其发展趋势，推动英语教育的革新。

当前，在线教学成为社会常态。英语学科的学习可以依据线上教育规律与特点，充分发挥平台辅助教学的优势，引领学生每天自主点读、跟读、预习相关内容，并及时通过平台判断学习的效果。学生对自己英语水平的评价从依赖教师有限的反馈进行判断转向对自身学习过程的数据分析。随着课堂教学内容、教学方式、学习过程的数字化发展，学生的教学评价也将逐渐数字化，对这些数据的分析和利用不仅限于教师，作为学习主体的学生也将参与评价的过程。这也体现了大数据时代教学方式与教学手段的新特点，每个学习主体都是某种意义上的中心，可以形成"反馈学习"机制，利用自己产生的数据进行调整，持续改善自己的表现。

作为教学的基本工具，教材是知识的载体，需充分体现教学内容和教学要求，促进教学质量提升。然而，目前国内现有的大学英语教材大多以提高学生的语言能力为目标，众多院校在培养应用型人才的过程中对匹配性教材的供给提出了新要求。在大数据英语教材建设方面，任务十分紧迫。在此关键时刻，《大数据专业英语》的问世可谓恰逢其时。

本书的目标在于使学生直接了解专业前沿知识和技术发展现状，以内容为依托（CBI），尽可能地发挥学生的主观能动性，让学生从注重"输入"转变为注重"输出"，提高学生的实践能力和对所学知识的应用能力。本书具有时代性、真实性，与专业设置接近，基于主题进行编排，全书共 8 个单元，涵盖了大数据的概念、特点、市场、技术、分析、应用、云计算、区块链、人工智能等。本书的教学层次清晰，每个单元都遵循同样的编排体系；在口语技能、阅读技能、翻译技能等方面都进行了较为合理的任务设计；提供教学重点和专业词汇注释；设置基于内容的阅读理解、词汇练习和翻译练习，使学生掌握大数据英语阅读、翻译技巧，扩充学生的专业英语词汇量；注重培养学生阅读大数据英语文章的能力，使学生能够掌握本学科的前沿知识，提高表达和思辨能力。

本书是为高等院校相关专业（尤其是数据科学与大数据技术、大数据管理与应用、信息管理、经济管理等专业开设的大数据相关课程）设计并编写的具有丰富实践特色的教材，也可作为有一定实践经验的 IT 应用人员、管理人员的参考书，以及继续教育的教材和英语类的专业科技英语课程的选修教材。本书反映了我国英语教育在大数据专业领域的最新开发与研究进展。在此，特别感谢《大数据专业英语》编委会为本书定稿、修改及出版付出的辛劳和汗水！

戴炜栋

教育部社科委员会学部委员

上海外国语大学原校长、英语教授、博士生导师

教育部高等学校外语专业教学指导委员会原主任委员

推荐序三

目前，大数据行业处于上升期，大量的人才缺口使行业薪资持续走高，发展前景良好。这对于广大的大数据专业学习者来说无疑是个好消息。

党的十八届五中全会将大数据上升为国家战略。回顾过去几年的发展，可以将我国大数据的发展总结为："进步长足，基础渐厚；喧嚣已逝，理性回归；成果丰硕，短板仍在；势头强劲，前景光明。"

作为人口大国和制造大国，我国产生的数据量大，数据资源极为丰富。随着数字中国建设的推进，各行业的数据资源采集、应用能力不断提升，将积累更多的数据。

我国虽然在大数据应用领域取得了较大进展，但是仍需大力推进在基础理论、核心器件和算法软件等层面的发展。唯有提升全民对大数据的正确认识，使其具备用大数据思维认识和解决问题的基本素质和能力，才能积极防范大数据带来的新风险；唯有加快培养适应未来需求的合格人才，才能在数字经济时代形成国家的综合竞争力。因此，学好大数据英语，有助于深入了解大数据的特点、分类发展方向、代表人物、国内外研究现状，掌握大数据前沿核心技术。

《大数据专业英语》从 3 个维度深入浅出地介绍了大数据。第 1 个维度介绍了大数据技术的起源、发展、关键技术和未来发展趋势；第 2 个维度从业务和技术的角度出发，介绍了实际案例，帮助读者理解大数据的用途及技术的本质；第 3 个维度讲解了如何将大数据与前沿的云计算、区块链、人工智能技术等结合起来。

《大数据专业英语》全面展现了大数据理论、应用和技术，具有重要的学术意义和实用价值。

高志凯

耶鲁法学院法学博士

全球化智库（CCG）副主任

欧美同学会常务理事

耶鲁法学院中国协会会长

推荐序四

In today's tech infused world, there are certain topics that have been around now for several years, but while the underlying landscape has been shifting at pace, many of the major technologies continue to be misunderstood. Big Data is coming into its own now, as Big Data English exposes, thanks to decreased storage cost, increased processor speed and the variety of sources generating data. In this context, many of the traditional companies continue to struggle with their digital transformation programmes. The same can now even be said of certain pure players born at the turn of the 21st century who have failed to adapt to an ever-changing environment. Big Data is a capital element of the new wave of digital technologies, not least of which is Artificial Intelligence. Spawned by the incredible volume and accessibility of data, new business models, opportunities and ventures are cropping up everywhere. In this book, there is a thorough and honest review of how Big Data works, providing an adapted lexicon of the necessary terms as well as the challenges and chances that Big Data affords us. The book explores how Big Data is intricately related with many of the other new technologies as well as applications and opportunities for Big Data within specific industries, including the important areas of healthcare and education, as well as other fundamental sectors such as logistics and transportation. It's a very educational book, with useful exercises at the end of each chapter. For people who need to get their heads around Big Data, this book provides a solid and complete overview.

Minter Dial

Professional speaker and author of Heartificial Empathy

Futureproof and The Last Ring Home

自 序

大数据正在改变中国，改变一切传统企业，改变整个市场格局。我们的工作、生活、社交都与它息息相关。从移动、电信、阿里巴巴、百度到互联网的每个角落，它无所不在，带领我们从信息技术时代进入数据技术时代。大数据产业的高速发展，要求从业人员掌握新技术、新方法。新术语层出不穷，也对从业人员的专业英语提出了更高的要求。从业人员只有提高专业英语水平才能及时获得最新、最先进的专业知识。在未来，只有具备大数据相关技能并精通专业英语的人才能赢得竞争，成为职场中不可或缺的核心人才与领军人物。

本书是易懂、实用、全面、新潮的专业英语教材。本书的特点与优势如下。

（1）选材全面、通俗易懂，包括大数据概念、特点、重要性、市场、分析、应用、安全、Hadoop 和 NoSQL，以及与大数据紧密相关的云计算、区块链、人工智能等。本书的覆盖面广，书中的许多内容非常实用。这在专业英语教材中并不多见。

（2）思路清晰、结构完整，非常适合教学，与课堂教学的各环节紧密结合，方便备课、教学、复习和考试。每单元由 7 部分组成。

- 学习目标（Learning Goals）：概括本单元的重点和难点，使读者一目了然。
- 对话（Dialogue）：图文并茂，激发读者的阅读兴趣，内容由浅入深，符合认知规律。
- 课文（Reading）：选材广泛、风格多样，既符合实际，又有深度。难度恰到好处，能满足各层次英语学习者的需求。
- 生词、短语和句子（New Words & Phrases & Sentences）：给出课文中出现的生词、短语和句子，方便读者学习和积累。
- 习题（Exercises）：以课文练习为主，加深读者对课文的理解。
- 对话翻译（Translations for Dialogue）：加深读者对对话内容的理解。
- 课文翻译（Translations for Reading）：方便读者进行对照，提高翻译能力。

（3）内容新颖时尚，重点突出，详略得当。以大数据处理、储存、分析和应用为重点，将云计算、区块链、人工智能与大数据结合。这种创意和体系在专业英语中较为少见。数据可视化（Data Visualization）、量子计算（Quantum Computing）、深度学习（Deep Learning）等术语的出现，也使读者耳目一新。

（4）案例丰富，深入浅出。本书展示了谷歌、Facebook、Twitter等国外大数据先锋企业，以及阿里巴巴、腾讯、百度等国内大数据快车企业最具代表性的应用案例。

本书既可以作为大数据专业相关课程的教材、英语专业及计算机专业的选修教材、各类院校大数据和相关专业的参考书，也可以作为计算机从业人员或有志投身大数据领域的人才的自学书籍。

本书的部分文章选自互联网，作者不详，无法列入参考书目，在此向文章原作者表示感谢。由于笔者水平有限，书中难免出现不足之处，请各位读者予以指正。

高文宇　徐　亭
2020 年 10 月

目 录

Unit One　Big Data Introduction ··· 001

Unit Two　Big Data Market ·· 018

Unit Three　Big Data Technology Stack Processing and Storage ············ 033

Unit Four　Big Data Target Analytics ·· 056

Unit Five　Big Data Application ··· 086

Unit Six　Big Data and Cloud Computing ··· 110

Unit Seven　Big Data and Blockchain ··· 127

Unit Eight　Big Data and Artificial Intelligence ·································· 135

Appendix A　49例大数据术语 ··· 144

Appendix B　总词汇表 ··· 155

Appendix C　书后练习答案 ·· 169

Appendix D　参考资料 ··· 182

01 Unit One Big Data Introduction

1.1 Learning Goals

After learning this unit, you will be able to master the following knowledge:

1. Big Data is a new generation of technologies and architectures designed to extract value economically from very large volumes of a wide variety of data by enabling high-velocity capture, discovery and analysis.

2. The five characteristics of Big Data that can be used to help differentiate data categorized as "Big" from other forms of data: Volume, Velocity, Variety, Veracity, Value.

3. Big Data has the potential to provide new forms of competitive advantages for organizations.

1.2 Dialogue

Ben Zhang, manager of Hangzhou Angel Software Co., Ltd., is talking with Neil Millers, the marketing manager from SCC Company. They want to enter into business partnership with each other.

Neil: Good morning, I'm Neil Millers from Italy. This is my business card.

Zhang: Nice to meet you, Miss Millers. I am Ben Zhang, manager of Hangzhou Angel Software Co., Ltd.

(Shaking hands and exchanging their bussiness cards.)

Neil: I was very impressed by your products displayed at the Canton Fair. And we now avail ourselves of this opportunity to visit you with a view to establishing partnership with you.

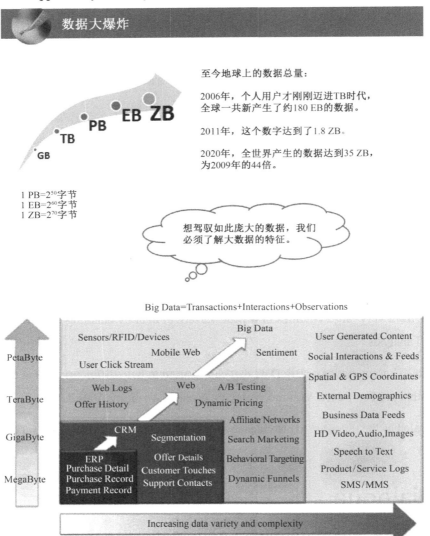

Zhang: Thanks. The samples displayed at the fair are just part of our products. We have been in the line of Big Data since 2010, and we enjoy a good reputation in the world market.

Neil: By the way, what does Big Data mean?

Zhang: Big Data is the term for a collection of data sets so large and complex that it is difficult to process them by using on-hand database management tools or traditional data processing applications.

Neil: How can we solve this problem?

Zhang: Big Data is usually transformed in three dimensions: volume, velocity, and variety.

Volume: Machine-generated data, which is produced in larger quantities than traditional data did.

Velocity: This refers to the speed of data processing.

Variety: This refers to a large variety of input data which in turn generates a large amount of data as output.

Neil: Can we capture and manage a lot of information at the same time?

Zhang: Of course. We can work with many new types of data, even bigger than TeraBytes, structured or unstructured. Meanwhile, we can exploit these masses of information and new data types with new styles of applications.

Neil: That sounds fantastic.

1.3 Reading

Recently, when people still feel vague about "Internet of Things" "Cloud Computing" "Mobile Internet" and other hot words, "Big Data" has emerged and developed into a prairie fire. The biggest difference between the 2014 Brazil World Cup and the previous World Cup is that it integrates many technological elements such as "Cloud Computing" "Big Data", etc. IBM research in 2013 shows that, 90% of all the data obtained by human civilization is generated in the past two years. By 2020, the data generated in the world is 44 times of that in 2009. According to IDC monitoring, the amount of data produced by human beings is growing exponentially, about doubling every two years, and the global amount reach 35 ZB in 2020. According to statistics, on average, 2 million users are using Google search every second. Facebook has more than 1 billion registered users and generates more than 300 TB of log data every day. At the same time, the rapid development of Sensor Networks, the Internet of Things, Social Networks and other technologies has led to the explosive growth of data scale. Various video monitoring and sensing devices have also continuously generated huge amount of streaming media data. Energy, transportation, health care, finance, retail and other industries have also generated a large number of data, accumulating TeraBytes and PetaBytes of Big

Data. The above situation shows that now it has entered the era of Big Data, which has begun to benefit mankind and become a valuable asset of the information society. A decade of digital universe growth is shown in Figure 1-1.

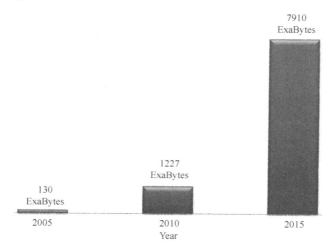

Figure 1-1   A decade of digital universe growth.

1.3.1 What is Big Data?

According to McKinsey, Big Data refers to data sets whose size is beyond the ability of typical database software tools to capture, store, manage and analyse. There is no explicit definition of how big a data set should be. New technology needs to be in place to manage this Big Data phenomenon. IDC defines Big Data technologies as a new generation of technologies and architectures designed to extract value economically from very large volumes of a wide variety of data by enabling high-velocity capture, discovery and analysis. According to O'Reilly, "Big Data is data that exceeds the processing capacity of conventional database systems. The data is too big, moves too fast, or does not fit the structures of existing database architectures. To gain value from these data, there must be an alternative way to process it."

1.3.2 Characteristics of Big Data

For a data set to be considered Big Data, it must possess one or more characteristics that require accommodation in the solution design and architecture of the analytic environment. Most of these characteristics were initially identified by Doug Laney in 2001 when he published an article describing the impact of the volume, velocity and variety of e-commerce data on enterprise data warehouses. To this list, veracity has been added to account for the lower signal-to-noise ratio of unstructured data as compared to structured data sources. Ultimately, the goal is to conduct analysis of the data in such a manner that high-quality results are delivered in a timely manner, which provides optimal value to the enterprise.

This section explores the five characteristics of Big Data that can be used to help differentiate data categorized as "Big" from other forms of data. The five Big Data characteristics shown in Figure 1-2 are commonly referred to as the "5V".

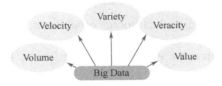

Figure 1-2　The "5V" of Big Data.

1. Volume

The anticipated volume of data that processed by Big Data solutions is substantial and ever-growing. High data volumes impose distinct data storage and processing demands, as well as additional data preparation, curation and management processes. Figure 1-3 provides a visual representation of the large volume of data created by organizations and users worldwide everyday.

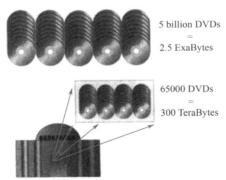

Figure 1-3　The large volume of data created by organizations and users worldwide everyday.

Typical data sources that are responsible for generating high data volumes can include as follows.

（1）Online transactions, such as point-of-sale and banking.

（2）Scientific and research experiments, such as the Large Hadron Collider and Atacama Large Millimeter or Submillimeter Array telescope.

(3) Sensors, such as GPS sensors, RFIDs, smart meters and telematics.

(4) Social media, such as Facebook and Twitter.

2. Velocity

In Big Data environments, data can arrive at fast speed, and enormous data sets can accumulate within very short periods of time. From an enterprise's point of view, the velocity of data translates into the amount of time it takes for the data to be processed once it enters the enterprise's perimeter. Coping with the fast inflow of data requires the enterprise to design highly elastic and available data processing solutions and corresponding data storage capabilities.

Depending on the data source, the velocity may be different. For example, MRI scan images are not generated as frequently as log entries from a high-traffic Web Server. As illustrated in Figure 1-4, data velocity is put into perspective when considering that the following data volume can easily be generated in a given minute: 350000 tweets, 300 hours of video footage uploaded to YouTube, 171 million emails and 330 GB of sensor data from a jet engine.

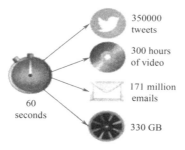

Figure 1-4　Examples of high-velocity Big Data data sets produced every minute include tweets, video, emails and GB generated from a jet engine.

3. Variety

Data variety refers to the multiple formats and types of data that need to be supported by Big Data solutions. Data variety brings challenges for enterprises in terms of data integration, transformation, processing, and storage. Figure 1-5 provides a visual representation of data variety, which includes structured data in the form of financial transactions, semi-structured data in the form of emails and unstructured data in the form of images.

Figure 1-5 A visual representation of data variety.

4. Veracity

Veracity refers to the quality or fidelity of data. Data that enters Big Data environments need to be assessed for quality, which can lead to data processing activities to resolve invalid data and remove noise. In relation to veracity, data can be part of the signal or noise of a data set. Noise is data that cannot be converted into information and thus has no value, whereas signals have value and lead to meaningful information. Data with a high signal-to-noise ratio has more veracity than data with a lower ratio. Data that is acquired in a controlled manner, for example via online customer registrations, usually contains less noise than data acquired via uncontrolled sources, such as blog postings. Thus the signal-to-noise ratio of data is dependent upon the source of the data and its type.

5. Value

Value is defined as the usefulness of data for an enterprise. The value characteristic is intuitively related to the veracity characteristic in that the higher the data fidelity, the more value it holds for the business. Value is also dependent on how long data processing takes, because analytics results have a shelf-life. For example, a 20 minutes delayed stock quote has little to no value for making a trade compared to a quote that is 20 milliseconds old. Data that has high veracity and can be analysed quickly has more value to a business, as shown in Figure 1-6. As demonstrated, value and time are inversely related. The longer it takes for data to be turned into meaningful information, the less value it has for a business. Stale results inhibit the quality and speed of decision making.

Apart from veracity and time, value is also impacted by the following lifecycle-related concerns.

(1) How well has the data been stored?

(2) Were valuable attributes of the data removed during data cleansing?

(3) Are the right types of questions being asked during data analysis?

(4) Are the results of the analysis being accurately communicated to the appropriate decision-makers?

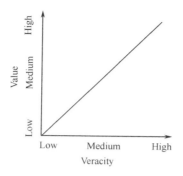 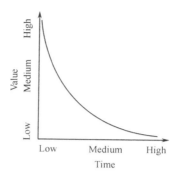

Figure 1-6　Data that has high veracity and can be analysed quickly has more value to a business.

1.3.3　Why is Big Data Important?

The convergence across business domains has ushered in a new economic system that is redefining relationships among producers, distributors, and consumers or goods and services. In an increasingly complex world, business verticals are intertwined and what happens in one vertical has a direct impact on other verticals. Within a business, this complexity makes it difficult for business leaders to rely solely on experience (or pure intuition) to make decisions. They need to rely on good data services for their decisions. By placing data at the heart of the business operations to provide access to new insights, organizations will then be able to compete more effectively.

Three things have come together to drive attention to Big Data.

(1) The technologies to combine and interrogate Big Data have matured to a point where their deployments are practical.

(2) The underlying cost of the infrastructure to power the analysis has fallen dramatically, making it economical to mine the information.

(3) The competitive pressure on businesses has increased to the point where most traditional strategies are offering only marginal benefits. Big Data has the potential to provide new forms of competitive advantage for businesses.

For years, organizations have captured structured transactional data and used batch processing to place summaries of the data into traditional relational database. The analysis of such data is retrospective and the investigations done on the data sets are on past patterns of business operations. In recent years, new technologies with lower costs have enabled improvements in data capture, data storage, and data analysis. Organizations can now capture more data from many more sources and types (blogs, social media, audio and video files). The options to optimally store and process the data have expanded dramatically and technologies such as MapReduce and in-memory computing (discussed in later sections) provide highly optimized capabilities for different business purposes. The analysis of data can be done in real-time, acting on full data sets rather than summarised elements. In addition, the number of options to interpret and analyse the data has also increased, with the use of various visualization technologies.

1.4 New Words & Phrases & Sentences

1.4.1 New Words

1. volume   *n.* 体积；容积；容量；量；额；音量；响度
2. variety   *n.* （同一事物的）不同种类，多种式样；变化；多样化
3. veracity   *n.* 真实；准确；诚实
4. inhibit   *v.* 阻止；阻碍；抑制；使拘束；使尴尬
5. interrogate   *v.* 讯问；审问；盘问；（在计算机或其他机器上）查询，询问
6. architecture *n.* 建筑学；建筑风格；架构
7. velocity   *n.* 速度，速率
8. analytics   *n.* 分析学，解析学，分析论
9. TeraByte   *n.* 太字节
10. fantastic   *adj.* 极好的；了不起的；很大的；大得难以置信的；怪诞的；荒诞不经的；富于想象的
11. accumulate   *v.* 积聚，堆积
12. ExaByte   *n.* 艾字节
13. phenomenon   *n.* 现象；杰出的人；非凡的人（或事物）
14. characteristic *adj.* 典型的；独特的；特有的

    n. 特征；特点；品质

1.4.2 Phrases

1. data warehouse   数据仓库
2. according to   根据
3. to be considered   待考虑
4. data cleansing   数据清洗
5. as well as   以及
6. accumulate within   在……范围内累积
7. be generated in   生成于
8. be supported by   由……支持
9. be defined as   被定义为
10. as demonstrated   如……所示
11. apart from   除了……
12. distributed computing system 分布式计算系统
13. makes it difficult for   使……很难
14. business operation   商业运营
15. drive attention   引起注意

1.4.3 Sentences

1. At the same time, the rapid development of Sensor Networks, the Internet of Things, Social Networks and other technologies has led to the explosive growth of data scale. 同时,传感器网络、物联网、社交网络等技术的快速发展导致数据规模出现爆发式增长。

2. Energy, transportation, health care, finance, retail and other industries have also generated a large number of data, accumulating TeraBytes and PetaBytes of Big Data. 能源、交通、医疗卫生、金融、零售等行业也有大量数据不断产生,积累了 TB 级、PB 级的大数据。

3. There is no explicit definition of how big a data set should be. New technology needs to be in place to manage this Big Data phenomenon. 还没有对大数据的数据集大小做出明确的定义。

必须采用新技术来管理这种大数据现象。

4. For a data set to be considered Big Data, it must possess one or more characteristics that require accommodation in the solution design and architecture of the analytic environment. 作为大数据的数据集必须具有一个或多个特征，这些特性需要在分析环境的解决方案设计和体系结构中进行调整。

5. Ultimately, the goal is to conduct analysis of the data in such a manner that high-quality results are delivered in a timely manner, which provides optimal value to the enterprise. 从根本上讲，目标是通过对数据进行分析的方式来及时交付能够为企业带来最佳价值的高质量结果。

6. The anticipated volume of data that processed by Big Data solutions is substantial and ever-growing. 大数据解决方案处理的预期数据量非常大且还在不断增长。

7. In Big Data environments, data can arrive at fast speed, and enormous data sets can accumulate within very short periods of time. 在大数据环境下，数据可以快速到达，巨大的数据集可以在很短的时间内积累起来。

8. Coping with the fast inflow of data requires the enterprise to design highly elastic and available data processing solutions and corresponding data storage capabilities. 为了应对快速的数据流入，企业需要设计高度灵活且可用的数据处理解决方案，并具备相应的数据存储能力。

9. Data that is acquired in a controlled manner, for example via online customer registrations, usually contains less noise than data acquired via uncontrolled sources, such as blog postings. 以受控方式（如通过在线客户注册）获取的数据通常比以非受控方式（如博客帖子）获取的数据包含的噪声更少。

10. Value is also dependent on how long data processing takes, because analytics results have a shelf-life. 因为分析结果存在保质期，所以价值还取决于处理数据所需的时间。

1.5　Exercises

【Ex. 1】Content Questions.

1. What is the definition of Big Data?

2. What are the characteristics of Big Data?

3. Why is Big Data important?

4. What is the value of Big Data?

【Ex. 2】句子翻译。

1. I hope that this talk has given you some insight into the kind of work that we've been doing.

2. The new systems have been optimized for running Microsoft Windows.

3. These designs demonstrate her unerring eye for colour and detail.

4. Let me make this clear: a bar chart is not analytics.

5. A good dictionary will give us the connotation of a word as well as its denotation.

6. The latest lifestyle trend is downshifting.

7. The end of an era presupposes the start of another.

8. You cannot combine structured and unstructured exception handling in the same function.

9. Finally, the practical application shows the feasibility and veracity of this approach.

10. The viability of multilayer switches depends on the protocol supported.

【Ex. 3】短文翻译。

1. Cloud Computing is a general term for anything that involves delivering hosted services over the Internet. These services are broadly divided into three categories: Infrastructure-as-a-Service (SaaS), Platform-as-a-Service (PaaS) and Software-as-a-Service (SaaS). The name Cloud Computing was inspired by the cloud symbol that's often used to represent the Internet in flowcharts and diagrams.

2. A cloud service has three distinct characteristics that differentiate it from traditional service. It is sold on demand, typically by the minute or the hour; it is elastic — a user can have as much or as little of a service as they want at any given time; and the service is fully managed by the provider (the consumer needs nothing but a personal computer and Internet access). Significant innovations in virtualization and distributed computing, as well as improved access to High-Speed Internet and a weak economy, have increased interest in Cloud Computing.

1.6 Translations for Dialogue

张本是杭州天使软件公司的经理。他正和SCC公司的销售部经理内尔·米尔斯交流。他们想在业务上建立合作关系。

内尔：早上好，我是来自意大利的内尔·米尔斯。这是我的名片。

张本：见到你很高兴，米尔斯小姐。我是张本，杭州天使软件公司的经理。

（握手并互换名片）

内尔：我对你们在广交会上的展品印象深刻。我们想借此机会拜访你们并与你们建立合作关系。

张本：谢谢。我们在展会上只展出了部分产品。2010年以来，我们一直聚焦于大数据领域，并在全球市场上享有盛誉。

内尔：顺便问一下，大数据是什么意思？

张本：大数据是数据的集合，这些数据集十分庞大和复杂，以至于使用现有的数据库管理工具或传统的数据处理应用程序很难对其进行处理。

内尔：我们如何处理这个问题呢？

张本：大数据通常在3个维度变化：量、速度和种类。

量：与传统数据相比，机器生成的数据量更多。

速度：数据处理的速度。

种类：输入数据的种类多样，且输入数据又会产生大量的输出数据。

内尔：我们能同时捕获和管理大量信息吗？

张本：当然。我们可以处理许多结构化或非结构化的新型数据，甚至大于TB级。同时，我们可以利用海量信息和新型数据来开发新的应用程序。

内尔：听起来棒极了。

1.7 Translations for Reading

近年来,当人们对"物联网""云计算""移动互联网"等热词的印象还很模糊时,"大数据"横空出世并发展成燎原之势。2014 年巴西世界杯与往届世界杯最大的区别在于其融入了"云计算""大数据"等众多技术元素。2013 年,IBM 的研究表明,在人类文明获得的所有数据中,有 90%是在过去两年内产生的。2020 年,全世界产生的数据达到 2009 年的 44 倍。IDC 的监测显示,人类产生的数据量呈指数增长,大约每两年翻一番,2020 年,全球数据量达到 35 ZB。据统计,平均每秒都有 200 万用户使用 Google 搜索。Facebook 有超过 10 亿的注册用户,每天生成 300 TB 以上的日志数据。同时,传感器网络、物联网、社交网络等技术的快速发展导致数据规模出现爆发式增长。各种视频监控和传感设备也源源不断地产生大量的流媒体数据,能源、交通、医疗卫生、金融、零售等行业也有大量数据不断产生,积累了 TB 级、PB 级的大数据。上述情况表明,现在已经进入大数据时代,大数据已经开始造福人类,并成为信息社会的宝贵财富。数字世界成长的 10 年如图 1-1 所示。

1.7.1 什么是大数据?

麦肯锡认为,大数据指的是规模超过典型数据库软件工具捕获、存储、管理和分析能力的数据集。还没有对大数据的数据集大小做出明确的定义。必须采用新技术来管理这种大数据现象。IDC 将大数据技术定义为新一代技术和体系结构,旨在通过实现高速捕获、发现和分析,从海量的数据中经济地获取价值。O'Reilly 指出,"大数据是指超过传统数据库系统处理能力的数据。这些数据太大、移动太快,或不适合现有数据库架构。要从这些数据中获得价值,必须采用另一种处理方法。"

1.7.2 大数据的特征

作为大数据的数据集必须具有一个或多个特征,这些特性需要在分析环境的解决方案设计和体系结构中进行调整。2001 年,Doug Laney 发表了一篇文章,描述了电子商务数据的量、速度和种类对企业数据仓库的影响,并初步确定了这些特征中的大多数。为了说明与结构化数据相比,非结构化数据的信噪比更低,还在此列表中增加了真实性。从根本上讲,目标是通过对数据进行分析的方式来及时交付能够为企业带来最佳价值的高质量结果。

本节探讨 5 个大数据特征,这些特征可用于帮助区分大数据与其他形式的数据。图 1-2 所示的 5 个大数据特征通常被称为"5V"。

1. 量

大数据解决方案处理的预期数据量非常大且还在不断增长。巨大的数据量带来了不同的数据存储和处理需求,以及额外的数据准备、监管和管理操作。图 1-3 直观地展示了全球组织和用户每天创建的大量数据。

负责生成大量数据的典型数据源如下。

(1)网上交易,如销售点和银行。

(2)大型强子对撞机、阿塔卡马大型毫米波或亚毫米波阵列望远镜等科研实验。

(3)传感器,如 GPS 传感器、RFID、智能仪表和远程信息处理。

(4)社交媒体,如 Facebook 和 Twitter。

2. 速度

在大数据环境下,数据可以快速到达,巨大的数据集可以在很短的时间内积累起来。从企业的角度来看,数据的速度代表了数据进入企业边界后,处理数据所花费的时间。为了应对快速的数据流入,企业需要设计高度灵活且可用的数据处理解决方案,并具备相应的数据存储能力。

数据传输速度可能随数据源的不同而变化。例如,MRI 扫描图像并没有高流量 Web 服务器的日志条目生成得频繁。如图 1-4 所示,当认为以下数据量可以在特定的一分钟内轻松生成时,才能正确地看待数据传输速度:350000 条推文、上传到 YouTube 的 300 小时录像片段、1.71 亿封电子邮件和来自喷气发动机的 330 GB 传感器数据。

3. 种类

数据种类是指大数据解决方案需要支持的多种数据格式和类型。数据种类为企业带来了数据集成、转换、处理和存储方面的挑战。数据种类的可视化表示如图 1-5 所示,包括金融交易形式的结构化数据、电子邮件形式的半结构化数据和图像形式的非结构化数据。

4. 真实性

真实性是指数据的质量或保真度。进入大数据环境的数据需要进行质量评估,可能会由数据处理机构去除无效数据和消除噪声。与真实性相关,数据可以是数据集信号或噪声的一部分。噪声是不能转换成信息的数据,因此,噪声没有价值,而信号有价值,能够通过它获得有意义的信息。与低信噪比数据相比，高信噪比数据具有更高的真实性。以受控方式（如通过在线客户注册）获取的数据通常比以非受控方式（如博客帖子）获取的数据包含的噪声更少。因此，数据的信噪比取决于数据的来源及类型。

5. 价值

价值是指数据对企业的作用。数据的价值特征与真实性特征有直接关系，即数据的保真度越高，它对企业的价值就越大。因为分析结果存在保质期，所以价值还取决于处理数据所需的时间。例如，与延迟 20 毫秒的股票报价相比，延迟 20 分钟的股票报价对交易来说几乎没有价值。具有高真实性强和分析速度快的数据对企业来说具有更高价值，如图 1-6 所示。从图中可以看出，价值与时间成反比。数据转化为有意义的信息所需的时间越长，它的价值就越小。过时的数据会影响决策的质量和速度。

除了真实性和时间，价值还受以下与生命周期相关的问题的影响。

（1）数据存储得有多好？

（2）在数据清洗期间是否删除了有价值的数据属性？

（3）在数据分析过程中提出的问题类型是否正确？

（4）分析结果是否能准确地传达给决策者？

1.7.3 为什么大数据很重要？

跨业务领域的融合带来了一种新的经济体系，该体系重新定义了生产者、分销商、消费者，以及商品和服务之间的关系。在一个越来越复杂的世界中，业务垂直领域是相互交织的，在一个垂直领域中发生的事情会直接影响其他垂直领域。在一家企业中，这种复杂性使企业领导者很难仅依靠经验（或纯粹的直觉）做出决策。他们需要依靠良好的数据服务来做出决策。通过将数据置于业务运营的核心位置来提供获取新见解的途径，企业能够进行更有效的竞争。

促使人们关注大数据的事件有 3 个。

（1）组合和查询大数据的技术已经成熟到可以实际部署的程度。

（2）为分析提供支持的基础设施的基本成本大幅下降，使信息挖掘的经济性增强。

（3）企业的竞争压力已经增加到大多数传统战略只提供边际效益的程度。大数据有潜力为企业提供新形式的竞争优势。

多年来，企业已经捕获了结构化交易数据，并使用批处理将数据汇总放入传统的关系型数据库中。对这些数据的分析是回顾性的，应针对过去的业务运营模式对数据集进行调查。近年来，成本较低的新技术使数据捕获、数据存储和数据分析得到了改进。企业现在可以从更多的来源和类型（博客、社交媒体、音频和视频文件）中获取更多数据。优化存储和处理数据的选项已经大量扩展，MapReduce 和内存计算（在后面的章节中讨论）等技术为不同业务目标提供了高度优化的功能。可以对数据进行实时分析，该分析作用于完整的数据集，而不是汇总的元素。此外，随着各种可视化技术的应用，用于解释和分析数据的选项数量也有所增加。

02 Unit Two Big Data Market

2.1 Learning Goals

After learning this unit, you will be able to master the following knowledge:

1. The convergence of mobile devices, the Mobile Internet and Social Networks provide an opportunity for organizations to derive competitive advantage through an efficient analysis of unstructured data.

2. Proliferation of the Internet of Things (IoT).

3. The "open source" nature of the Big Data technologies will encourage bigger adoption.

4. Many organizations perceive Big Data as an important development.

5. Data Visualization is both an art and a science, and has become an active area of research, teaching and development.

2.2 Dialogue

Jack Ma said that Alibaba is essentially a data company, and taobao's purpose is not to sell goods, but get all retail and manufacturing data.

Ben Zhang, manager of Hangzhou Angel Software Co., Ltd., is talking with Neil Millers, the marketing manager from SCC Company. They are talking about Big Data products and opportunities.

Zhang: Today, I would like to tell you that Big Data is useful. We can use Big Data to understand customers' favorites and meet their demands.

Neil: That's fine.

Zhang: As a well-known enterprise, our company specializes in R&D, manufacturing and marketing of Big Data products.

Neil: Good. But I need to study your products, brochure and price list further.

Zhang: No problem. I will send email with attachments this afternoon. Big Data is improving our life. Look at our smart watches or smart bracelets. They can generate the latest data, which tells us our calorie consumption and sleep quality.

Neil: Can they predict disease?

Zhang: Of course. Last month, the subsidiary of Alibaba Group Holding Ltd. wheeled out its cloud platform in cooperation with Wanli Cloud Medical Information Technology Co., Ltd. in Beijing. AI system has been built into the platform. Known as "Doctor You", it can help medical professionals with clinical diagnosis. And with data analysis, scientists can even decode the whole DNA within several minutes.

Neill: Quite impressive! I hope we can keep in touch and build long-term business relationship in the future.

Zhang: Thank you very much.

Neil: It is my pleasure. See you.

Zhang: See you.

2.3 Reading

As the vendor ecosystem around Big Data matures and users begin exploring more strategic business use cases, the potential of Big Data's impact on data management and business analytics initiatives will grow significantly. According to IDC, the Big Data technology and service market was about US$4.8 billion in 2011. The market size was about US$16.9 billion in 2015 as shown in Figure 2-1.

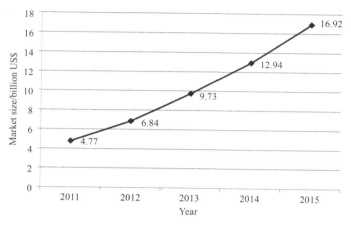

Figure 2-1   Global Big Data market projection.

China's Big Data sector will continue to post steady expansion by 2023, driven by policy support and technology integration, said a report from the International Data Corporation (IDC).

By 2023, the size of China's Big Data market is estimated to hit US$22.49 billion.

With the soaring of data, IDC predicts that Artificial Intelligence software platform will become the third-largest sub-market of the overall Big Data industry by 2023, replacing commercial services.

There are several factors that will drive this market.

2.3.1　Continuous Growth of Digital Content

The increasing market adoption of mobile devices that are cheaper, more powerful and packed with more Apps and functionalities is a major driver of the continuing growth of unstructured data. By 2020, the number of smartphone in the market reach 1.82 billion. The market adoption of tablets is also expected to increase significantly over the next few years, further contributing to the growth of data. In 2012, the shipment of tablets is 118.9 million. By 2020, the number rise to 3.4 billion. This market adoption of mobile devices and the prevalence of the Mobile Internet will see consumers increasingly being connected, using social media networks as their communication platform as well as their source of information.

The convergence of mobile devices, the Mobile Internet and Social Networks provide an opportunity for businesses to derive competitive advantage through an efficient analysis of unstructured data. Businesses that were early adopters of Big Data technologies and that based their business on data-driven decision making were able to achieve greater productivity of up to 5% or 6% higher than the norm. Big Data technology early adopters such as Facebook, LinkedIn, Walmart and Amazon are good examples for companies that plan to deploy Big Data analytics.

2.3.2 Proliferation of the Internet of Things (IoT)

According to Cisco's Internet Business Solutions Group (IBSG), 50 billion devices have been connected to the Internet by 2020. Meanwhile, Gartner reported that more than 65 billion devices were connected to the Internet by 2010. By 2020, this number goes up to 230 billion. These Internet-connected devices, ranging from smart meters to a wide range of sensors and actuators continually send out huge amount of data that need to be stored and analysed. Companies that deploy Sensor Networks will have to adopt relevant Big Data technologies to process a large amount of data sent by these networks. The growth of the number of Internet-connected devices is shown in Figure 2-2.

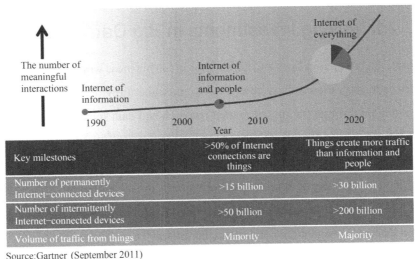

Source: Gartner (September 2011)

Figure 2-2 Exponential growth in the number of Internet-connected devices.

2.3.3 Strong Open Source Initiatives

Many of the technologies within the Big Data ecosystem have an open source origin, due to participation, innovation and sharing by commercial providers in open source development projects. The Hadoop framework, in conjunction with additional software components such as the open source R language and a range of open source database (Not only Structured Query Language database such as Cassandra and Apache HBase). The popularity and viability of these open source tools have driven vendors to launch their own open source tools (e.g., Oracle's version of the NoSQL database) or integrate these tools with their products.

Some of the technology companies that driving the technology evolution of Big Data are affiliated to the open source community in different ways. For example, Cloudera is an active contributor to various open source projects while EMC's Greenplum launched its Chorus as an open source tool to enable collaboration on data sets in a Facebook-like way. Hortonworks has also formed a partnership with Talend to bring the world's most popular open source data integration platform to the Apache community. The situation that open source technologies dominate the Big Data solutions may perpetuate as the technologies are changing rapidly and the technology standards are not well established. In turn, this poses a significant risk to any vendors who want to invest in proprietary Big Data technologies. Hence, the "open source" nature of the Big Data technologies will encourage bigger adoption.

2.3.4　Increasing Investments in Big Data Technologies

The information has always been a differentiator in the business world, allowing better business decisions to be made in an increasingly competitive landscape. Previously, market information was largely made available through traditional market research and data specialists. Today, virtually any company with large data sets can potentially become a serious player in the new information game. The value of Big Data will become more apparent to corporate leadership as companies seek to become more "data-driven" organizations. According to O'Reilly, a data-driven organization is one that "acquires, processes and leverages data in a timely fashion to create efficiency, iterate on and develop new products to navigate the competitive landscape."

The Big Data Insight Group survey of 300 senior personnel from a broad range of industry sectors revealed that many organizations are seeing Big Data as an important area for their organizations. The results is shown in Figure 2-3, among the respondents, 50% indicated current research into and sourcing of Big Data solutions while another 33% acknowledged that they were implementing or had implemented some form of Big Data solutions. This survey indicates that many organizations perceive Big Data as an important development and this interest could translate into future demand for Big Data technologies.

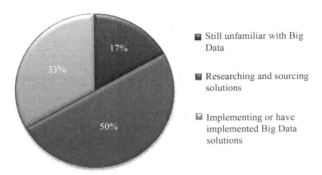

Figure 2-3　The Big Data Insight Group survey results.

2.3.5　Data Visualization Driven by the Information-based Economy

Data Visualization is both an art and a science. The rate at which data is generated has increased, driven by an increasingly information-based economy. Data created by activities of the Internet and an expanding number of sensors in the environment, such as satellites and traffic cameras, are referred to as "Big Data". Processing, analysing and communicating this data presents a variety of ethical and analytical challenges for Data Visualization. The field of data science and practitioners called data scientists have emerged to help address this challenge.

Data Visualization refers to the techniques used to communicate data or information by encoding it as visual objects (e.g., points, lines or bars) contained in graphics. It is one of the steps in data analysis or data science. According to Friedman, the main goal of Data Visualization is to communicate information clearly and effectively through graphical means. It doesn't mean that Data Visualization needs to look boring to be functional or extremely sophisticated to look beautiful. To convey ideas effectively, both aesthetic form and functionality need to go hand in hand, providing insights into a rather sparse and complex data set by communicating its key-aspects in a more intuitive way. An ideal visualization should not only communicate clearly, but stimulate viewer' engagement and attention. Well-crafted Data Visualization helps uncover trends, realize insights, explore sources, and tell stories.

Data Visualization is closely related to information graphics, information visualization, scientific visualization, exploratory data analysis and statistical graphics. In the 2/st century, Data Visualization has become an active area of research, teaching and development.

2.4 New Words & Phrases & Sentences

2.4.1 New Words

1. convergence　　　　　　　　n.（不同思想、群体或社会的）趋同，融合
2. proliferation　　　　　　　　n. 激增；涌现；增殖；大量的事物
3. subsidiary　　　　　　　　　adj. 辅助的；附带的；次要的；附属的；隶属的
　　　　　　　　　　　　　　　n. 附属公司；子公司
4. device　　　　　　　　　　　n. 装置；仪器；器具；设备
5. initiative　　　　　　　　　　n. 倡议；新方案；主动性；积极性；自发性
6. framework　　　　　　　　　n. 框架，结构
7. indicate　　　　　　　　　　v. 表明；显示；象征；暗示；间接提及；示意
8. perceive　　　　　　　　　　v. 注意到；意识到；察觉到；将……理解为；将……视为；认为
9. congruent　　　　　　　　　adj.（与 with 连用）一致的，适合的
10. visualization　　　　　　　　n. 显示，直观性；形象表示；显色
11. landscape　　　　　　　　　n. 风景，景色；乡村风景画；乡村风景画的风格；（文件的）横向打印格式；对……做景观美化
12. unstructured　　　　　　　　adj. 非结构化的
13. brochure　　　　　　　　　　n. 资料（或广告）手册
14. ecosystem　　　　　　　　　n. 生态系统
15. conjunction　　　　　　　　n. 连词，结合，同时发生

2.4.2 Phrases

1. proliferation of the Internet of Things　　物联网的普及
2. data-driven organization　　数据驱动型企业
3. continuous growth of digital content　　数字内容的持续增长
4. increase significantly　　显著增加
5. be expected to　　期望；预计；有望做某事
6. communication platform　　通信平台
7. mobile device　　移动设备
8. send out　　发送
9. strong open source initiatives　　强大的开源计划
10. due to　　由于
11. the technology evolution　　技术演进
12. an increasingly competitive landscape　　竞争日益激烈的局面
13. Data Visualization　　数据可视化
14. seek to　　寻求
15. data integration　　数据集成

2.4.3 Sentences

1. The increasing market adoption of mobile devices that are cheaper, more powerful and packed with Apps and functionalities is a major driver of the continuing growth of unstructured data. 越来越多的市场采用更便宜、功能更强大、包含更多应用程序和功能的移动设备，这是非结构化数据持续增长的主要动力。

2. This market adoption of mobile devices and the prevalence of the Mobile Internet will see consumers increasingly being connected, using social media networks as their communication platform as well as their source of information. 移动设备的市场应用和移动互联网的普及将使消费者的联系越来越紧密，并将社交媒体网络作为他们的通信平台和信息源。

3. The convergence of mobile devices, the Mobile Internet and Social Networks provide an opportunity for businesses to derive competitive advantage through an efficient analysis of unstructured data. 移动设备、移动互联网和社交网络的融合为企业提供了一个通过有效分析非结构化数据来获得竞争优势的机会。

4. Businesses that were early adopters of Big Data technologies and that based their business on data-driven decision making were able to achieve greater productivity of up to 5% or 6% higher than the norm. 早期采用大数据技术的企业和基于数据驱动制定业务决策的企业能够实现比常规企业高 5%或 6%的生产率。

5. These Internet-connected devices, ranging from smart meters to a wide range of sensors and actuators continually send out huge amount of data that need to be stored and analysed. 这些联网设备（从智能仪表到各种各样的传感器和执行器）不断发送大量需要被存储和分析的数据。

6. Companies that deploy Sensor Networks will have to adopt relevant Big Data technologies to process a large amount of data sent by these networks. 部署传感器网络的公司将不得不采用相关的大数据技术来处理这些网络发送的大量数据。

7. Some of the technology companies that driving the technology evolution of Big Data are affiliated to the open source community in different ways. 一些驱动大数据技术演进的科技企业以不同的方式隶属于开源社区。

8. The situation that open source technologies dominate the Big Data solutions may perpetuate as the technologies are changing rapidly and the technology standards are not well established. 由于技术日新月异且技术标准不完善，开源技术在大数据解决方案中占主导地位的情况可能会长期存在。

9. The information has always been a differentiator in the business world, allowing better business decisions to be made in an increasingly competitive landscape. 信息一直是商业世界中的独特因素之一，可以利用它在竞争日益激烈的局面下做出更好的商业决策。

10. The value of Big Data will become more apparent to corporate leadership as companies seek to become more "data-driven" organizations. 随着企业力图成为更大的"数据驱动"型企业，对于企业领导层来说，大数据的价值将变得更加明显。

2.5 Exercises

【Ex. 1】 **Content Questions.**

1. How many factors will drive the market?

2. What are the advantages of Big Data technology?

3. What is the core of Big Data?

4. What is the value of Big Data?

【Ex. 2】判断以下叙述的正误。

1. A database is a collection of organized information.

2. A relational database is a tabular database in which data is defined so that it can be reorganized and accessed in a number of different ways.

3. A distributed database is one that can be dispersed or replicated at certain points in a network.

4. An object-oriented programming database is one that is congruent with the data defined in object classes and subclasses.

5. Databases and database managers are prevalent only in large mainframe systems.

6. The most typical DBMS is a distributed Database Management System.

7. A DBMS can be thought of like a file manager that manages data in databases.

8. The records make up the columns and the fields make up the table rows.

【Ex.3】选择填空。

 (1) the analysis emphasizes the drawing of pictorial system models to document and validate both existing and proposed systems. Ultimately, the system models become the (2) for designing and constructing an improved system. (3) is such a technique. The emphasis in this technique is process-centered. Systems analysts draw a series of process models called (4) . (5) is another such technique that integrates data and process concerns into constructs called objects.

(1) A. Prototyping   B. Accelerated   C. Model-driven   D. Iterative

（2）A. image　　　　　B. picture　　　　C. layout　　　　D. blueprint

（3）A. Structured Analysis　　　　　B. Information Engineering
　　C. Discovery prototyping　　　　D. Object-Oriented Analysis

（4）A. PERT　　　　B. DFD　　　　C. ERD　　　　D. UML

（5）A. Structured Analysis　　　　　B. Information Engineering
　　C. Discovery prototyping　　　　D. Object-Oriented Analysis

【Ex. 4】根据给出的汉语词义和规定的词类写出相应的英语单词，每个单词的首字母已给出。

1. *vt.* 用公式表示，规则化　　　　　f_____
2. *adj.* 多余的，冗余的　　　　　　　r_____
3. *v.* 展开，配置　　　　　　　　　　d_____
4. *adj.* 离散的，分散的　　　　　　　s_____
5. *n.* 批　　　　　　　　　　　　　　b_____
6. *n.* 服务器　　　　　　　　　　　　s_____
7. *vi.* 改变，转化，变换　　　　　　t_____
8. *adj.* 不一致的，不协调的，矛盾的　i_____
9. *n.* 式样，模式　　　　　　　　　　p_____
10. *n.* 联机，在线式　　　　　　　　o_____
11. *n.* 知识库，仓库　　　　　　　　r_____
12. *adj.* 全异的　　　　　　　　　　d_____
13. *adj.* 错误的，不准确的　　　　　i_____
14. *adj.* 基础的，基本的　　　　　　f_____
15. *adj.* 多维的　　　　　　　　　　m_____
16. *n.* 指数　　　　　　　　　　　　e_____
17. *adj.* 健壮的　　　　　　　　　　r_____

18. *n.* 功能性　　　　　　　　　　　　f_____

19. *adj.* 可升级的　　　　　　　　　　s_____

20. *n.* 执行　　　　　　　　　　　　　i_____

【Ex.5】句子翻译。

1. You can chat with other people online.

2. Each summer a new batch of students try to find work.

3. Are your products and services competitive? How about marketing?

4. My server went wrong this morning.

5. It is proverbially easier to destroy than to construct.

6. The photochemical reactions transform the light into electrical impulses.

7. Experimental results show that the algorithm is robust to resist normal and geometrical attacks.

8. The feedback that comes from disparate industries and different areas have different results.

9. The other fundamental consideration in the conception of a plan is the function.

10. Populations tend to grow at an exponential rate.

2.6　Translations for Dialogue

张本是杭州天使软件公司的经理。他正在和 SCC 公司的销售部经理内尔·米尔斯谈论大数据产品及机遇。

张本：今天，我想告诉你大数据是有用的。我们可以利用大数据来了解客户的喜好，并满足他们的需求。

内尔：好的。

张本：我们公司是一家专业从事大数据产品研发、生产、销售的知名企业。

内尔：很好。但我需要进一步研究你们的产品、手册和价目表。

张本：没问题。今天下午我会发一封带有附件的电子邮件。大数据正在改善我们的生活。我们的智能手表或智能手环可以生成最新的数据，检测我们消耗的卡路里和睡眠质量。

内尔：它们可以预测疾病吗？

张本：当然。上个月，阿里巴巴集团控股有限公司的子公司与北京万里云医疗信息科技有限公司合作推出了云平台。该平台内置了人工智能系统，被称为"您的医生"，可以帮助医学专业人员进行临床诊断。通过数据分析，科学家甚至可以在几分钟内解码整个 DNA。

内尔：真令人印象深刻！我希望我们今后能保持联系并建立长期的业务关系。

张本：非常感谢。

内尔：不客气，再见。

张本：再见。

2.7 Translations for Reading

随着围绕大数据的供应商生态系统的成熟，用户开始探索更具战略性的业务用例，大数据对数据管理和商业分析举措的影响潜力将显著增长。IDC 的数据显示，2011 年，大数据技术和服务市场规模约为 48 亿美元，2015 年，市场规模约为 169 亿美元，如图 2-1 所示。

国际数据公司（IDC）的一份报告称，在政策支持和技术整合的推动下，到 2023 年中国大数据行业将继续稳步扩张。

预计 2023 年，中国大数据市场规模达到 224.9 亿美元。

随着数据的迅速增加，IDC 预测，2023 年人工智能软件平台将取代商业服务，成为整个大数据行业的第 3 大分市场。

以下因素将推动该市场的发展。

2.7.1 数字内容的持续增长

越来越多的市场采用价格更便宜、功能更强大、包含更多应用程序和功能的移动设备，这是非结构化数据持续增长的主要动力。2020 年，市场上的智能手机数量达到 18.2 亿台。预计未来几年市场采用的平板电脑也将显著增加，进一步促进数据的增长。2012 年，平板电脑的出货量达到 1.189 亿台；2020 年增至 34 亿台。移动设备的市场应用和移动互联网的普及将使消费者的联系越来越紧密，并将社交媒体网络作为他们的通信平台和信息源。

移动设备、移动互联网和社交网络的融合为企业提供了一个通过有效分析非结构化数据来获得竞争优势的机会。早期采用大数据技术的企业和基于数据驱动制定业务决策的企业能够实现比常规企业高5%或6%的生产率。对于计划部署大数据分析的企业来说，Facebook、LinkedIn、沃尔玛和亚马逊等早期采用大数据技术的企业是很好的参考。

2.7.2 物联网的扩散

思科互联网商务解决方案集团（IBSG）的数据显示，2020年，有500亿台设备接入互联网。同时，Gartner的报告指出，2010年，互联网接入设备超过650亿台；2020年，达到2300亿台。这些联网设备（从智能仪表到各种各样的传感器和执行器）不断发送大量需要被存储和分析的数据。部署传感器网络的公司将不得不采用相关的大数据技术来处理这些网络发送的大量数据。联网设备的数量增长情况如图2-2所示。

2.7.3 强大的开源计划

由于商业供应商参与、创新和共享开源开发项目，大数据生态系统中的许多技术都是开源的。Hadoop框架与其他软件组件（如开源R语言）和一系列开源数据库（NoSQL数据库，如Cassandra和Apache HBase）结合。这些开源工具的流行和可行性促使供应商推出自己的开源工具（如Oracle版本的NoSQL数据库）或将这些工具与其产品集成。

一些驱动大数据技术演进的科技企业以不同的方式隶属于开源社区。例如，Cloudera是各种开源项目的积极贡献者，EMC的Greenplum将它的Chorus作为一种开源工具启用，以类似Facebook的方式在数据集上实现协作。Hortonworks还与Talend建立了合作关系，将世界上最流行的开源数据集成平台引入Apache社区。由于技术日新月异且技术标准不完善，开源技术在大数据解决方案中占主导地位的情况可能会随技术的变化和技术标准的建立而长期存在。反过来，这会为任何想投资专有大数据技术的供应商带来重大风险。因此，大数据技术的"开源"性质将使其采用率更高。

2.7.4 增加对大数据技术的投资

信息一直是商业世界中的独特因素之一，可以利用它在竞争日益激烈的局面下做出更好的商业决策。过去，市场信息主要通过传统的市场研究和数据专家获得。如今，几乎任何拥有大型数据集的企业都有可能成为新信息游戏的重要参与者。随着企业力图成为更大的"数据驱动"型企业，对于企业领导层来说，大数据的价值将变得更加明显。O'Reilly认为，数据驱动型企业是"及时获取、处理和利用数据以提高效率，迭代和开发新产品以确定竞争格局"的企业。

Big Data Insight Group 对来自各行各业的 300 名高级人员的调查显示，许多企业都将大数据视为其重要领域。调查结果如图 2-3 所示，在受访者中，有 50%的人表示目前正在研究和寻求大数据解决方案，另有 33%的人表示他们正在实施或已经实施了某种形式的大数据解决方案。这项调查表明，许多企业认为大数据是一项重要的发展，这些企业对大数据的关注可能转化为未来对大数据技术的需求。

2.7.5 信息化经济驱动下的数据可视化

数据可视化是一门艺术，也是一门科学。在日益信息化的经济的驱动下，数据的生成速度加快。互联网活动产生的数据及环境中越来越多的传感器（如卫星和交通摄像机）被称为"大数据"。处理、分析和交流这些数据对数据可视化提出了各种伦理和分析挑战。数据科学领域和被称为数据科学家的实践者已经出现，并将应对这一挑战。

数据可视化是指通过将数据或信息编码为图像中包含的可视对象（如点、线或条）来进行数据或信息交流的技术。它是数据分析或数据科学中的步骤之一。Friedman 认为，数据可视化的主要目标是通过图像方式清晰、有效地交流信息。这并不意味着数据可视化需要看起来枯燥乏味才能实现功能，也不意味着需要非常复杂才能看起来精致。为了有效地传达想法，美学形式和功能需要齐头并进，通过以更直观的方式传达其关键方面来为稀疏和复杂的数据集提供见解。理想的可视化不仅要清楚地传达信息，还要刺激观众的参与和关注。精心设计的数据可视化有助于发现趋势、认识见解、探索来源和讲述故事。

数据可视化与信息图形学、信息可视化、科学可视化、探索性数据分析、统计图形学密切相关。在 21 世纪，数据可视化已经成为研究、教学和开发的活跃领域。

Unit Three Big Data Technology Stack Processing and Storage

3.1 Learning Goals

After learning this unit, you will be able to master the following knowledge:

1. Big Data technology can be broken down into two major components: the hardware component and the software component. The hardware component refers to the component and infrastructure layer. The software component can be further divided into data organization and management software, analytics and discovery software, decision support and automation software.

2. Big Data sets do not naturally eliminate data bias. The data collected can still be incomplete and distorted which, in turn, can lead to skewed conclusions. Big Data can also raise privacy concerns and reveal unintended information.

3. Hadoop is a framework that provides open source libraries for distributed computing using MapReduce software and its own Distributed File System, simply known as the Hadoop Distributed File System (HDFS).

4. The HDFS is a fault-tolerant storage system that can store huge amount of information, scales up incrementally and survives storage failure without losing data.

5. In simple terms, MapReduce is a programming paradigm that involves distributing a task across multiple nodes running a "map" function.

6. In addition to MapReduce and HDFS, Hadoop also refers to a collection of other software projects that use the MapReduce and HDFS framework.

7. Semi-structured data set and unstructured data set are the two fastest-growing data types in the digital universe. Hadoop HDFS and MapReduce enable the analysis of these two data types.

8. NoSQL (Not Only Structured Query Language) as a prime representative of the cloud data management system is widely applied. There are three popular types of NoSQL database: key-value store, columnar store, document store.

9. NoSQL database generally process data faster than relational database but confronted with the following challenges: overhead and complexity, reliability, consistency, unfamiliarity with the technology, limited ecostructure.

3.2 Dialogue

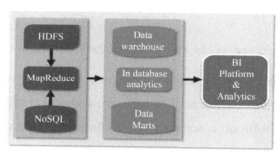

Storing and processing Big Data with the Hadoop framework

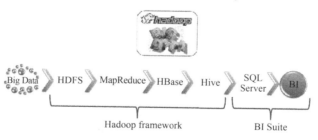

Big Data is becoming more and more popular, Tom would like to know more about it.

Tom: Is Big Data a challenge?

Mike: Exactly! Big Data is a relatively new phenomenon that businesses are just beginning to deal with. The challenge is to organize the data as it arrives, processes it if necessary, and store it where it can be easily retrieved.

Tom: Sounds very interesting.

Mike: Yes. Regarding the variety, Big Data commonly consists of multiple data types such as texts, images, videos, numbers, and audios that are generated from transaction processing systems, sensors, social media, smartphones and a diversity of other sources.

Jenny: However, some data may be structured, while other data is unstructured. Organizing this mix of data is challenging and may require database designers to seek solutions other than the rigid structure of relational database.

Tom: You said it. By the way, how about low-density data?

Mike: Low-density data refers to a large volume of very detailed data in which many of the details are not important. The opposite is high-density data, which is packed with lots of useful information.

Tom: I see! Anyway, thanks for your time.

3.3 Reading

3.3.1 Big Data Technology Stack

There are no comprehensive Big Data technology standards in place today. The main reason is that the Big Data analytics projects companies taking on are typically complex and diverse in nature.

Hadoop is almost synonymous with the term "Big Data" in the industry and is popular for handling a huge volume of unstructured data. The Hadoop Distributed File System enables a highly scalable, redundant data storage and processing environment that can be used to execute different types of large-scale computing projects.

Big Data technology stack can be broken down into two major components: the hardware component and the software component, as shown in Figure 3-1. The hardware component refers to the infrastructure. The software component can be further divided into data organization and management software, data analytics and discovery software, decision support and automation software.

1. Infrastructure

Infrastructure is the foundation of the Big Data technology stack. The main components of any data storage infrastructure (industry standard x86 servers and networking bandwidth of 10 Gbps) may be extended to a Big Data storage facility. Storage systems are also becoming more flexible and are being designed in a scale-out fashion, enabling the scaling of system performance and capacity. In-memory computing is supported by increased capabilities in system memory delivered at lower prices, making multi-GigaBytes (even multi-TeraBytes) memory more affordable.

Figure 3-1 Big Data technology stack.

2. Data Organization and Management Software

This layer refers to the software that processes and prepares all types of structured and unstructured data for analysis. This layer extracts, cleanses, normalizes and integrates data. Two Architectures — the extended Relational Database Management System (RDBMS) and the NoSQL Database Management System — have been developed to manage the different types of data.

The extended RDBMS is optimized for scale and speed in processing huge relational data (i.e., structured data) sets, adopting approaches such as using columnar store to reduce the number of table scans (columnar database) and exploiting Massively Parallel Processing (MPP) frameworks. Moreover, the NoSQL Database Management System (NoSQL DBMS) grew out of the realization that SQL's transactional qualities and detailed indexing are not suitable for the processing of unstructured files.

3. Data Analytics and Discovery Software

This layer comprises two data analytics software segments — software that supports offline, ad hoc, discovery and deep analytics, and software that supports dynamic real-time analysis and automated, rule-based transactional decision making. The tools can also be categorized by the type of data being analysed, such as text, audio and video. The tools within this layer are at different levels of sophistication. There are tools that allow for highly complex and predictive analysis as well as tools that simply help with basic data aggregation and trend reporting. In any case, the usage of the tools is not mutually exclusive — there can be a set of tools with different features residing in a system to enable Big Data analytics.

4. Decision Support and Automation Software

The process of data analysis usually involves a closed-loop decision making model, which at the minimum, includes steps such as track, analyse, decide and act. To support decision making and to ensure that action is taken, based on data analysis, is not a trivial matter. From a technology perspective, additional functionalities such as decision capture and retention are required to support collaboration and risk management.

There are two decision support and automation software categories: transactional decision management software and project-based decision management software. The former is automated, embedded within applications, real-time and rules-based in nature. It enables the use of outputs to prescribe or enforce rules, methods and processes. Examples include fraud detection, securities trading, airline pricing optimization, product recommendation and network monitoring. Project-based decision management software is typically standalone, ad hoc and exploratory in nature. It can be used for forecasting and estimation of trends. Examples include applications for customer segmentation for targeted marketing, product development and weather forecasting.

3.3.2 Hadoop MapReduce and Hadoop Distributed File System (HDFS)

Hadoop is a framework that provides open source libraries for distributed computing using MapReduce software and its own Distributed File System, simply known as the Hadoop Distributed File System (HDFS). It is designed to scale-out from a few computing nodes to thousands of machines, each offering local computation and storage. One of Hadoop's main value propositions is that it is designed to run on commodity hardware such as commodity servers or personal computers, and has a high tolerance for hardware failure. In Hadoop, hardware failure is treated as a rule rather than an exception.

1. HDFS

HDFS is a fault-tolerant storage system that can store huge amount of information, scales up incrementally and survives storage failure without losing data. Hadoop clusters are built with inexpensive computers. If one computer (or node) fails, the cluster can continue to operate without losing data or interrupting work by simply redistributing the work to the remaining machines in the cluster. HDFS manages storage on the cluster by breaking files into small blocks and storing duplicated copies of them across the pool of nodes. Illustration of distributed file storage using HDFS is shown in Figure 3-2. In this example, the entire data set will still be available even if two of the servers have failed.

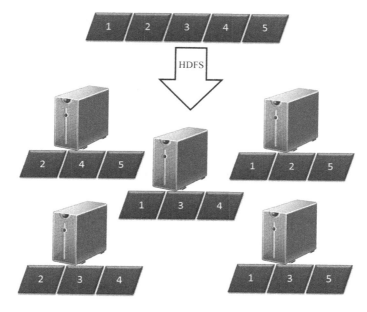

Figure 3-2　Illustration of distributed file storage using HDFS.

Compared to other redundancy techniques, including the strategies employed by Redundant Array of Independent Disks (RAID) machines, HDFS offers two key advantages. Firstly, HDFS requires no special hardware as it can be built from common hardware. Secondly, it enables an efficient technique of data processing in the form of MapReduce.

2. MapReduce

Most enterprise data management tools (Database Management Systems) are designed to make simple queries run quickly. Typically, the data is indexed so that only small portions of the data need to be examined in order to answer a query. This solution, however, does not work for data that cannot be indexed, namely the semi-structured data (text files) or unstructured data (media files). To answer a query in this case, all the data has to be examined. Hadoop uses the MapReduce technique to carry out this exhaustive analysis quickly. MapReduce is a data processing algorithm that uses a parallel programming implementation. In simple terms, MapReduce is a programming paradigm that involves distributing a task across multiple nodes running a "map" function. The map function takes the problem, splits it into subparts and sends them to different machines so that all the subparts can run concurrently. The results from the parallel map functions are collected and distributed to a set of servers running "reduce" functions, which then takes the results from the subparts and recombines them to get the single answer.

3. The Hadoop Ecosystems

In addition to MapReduce and HDFS, Hadoop also refers to a collection of other software projects that uses the MapReduce and HDFS framework. Table 3-1 briefly describes some of these tools.

Table 3-1 Some software tools

HBase	A key-value Database Management System that runs on HDFS
Hive	A system of functions that support data summarization and ad hoc query of the Hadoop MapReduce result set used for data warehousing
Pig	High-level language for managing data flow and application execution in the Hadoop environment
Mahout	A machine-learning system implemented on Hadoop
Zookeeper	Centralized service for maintaining configuration information, naming, providing distributed synchronization and group services
Sqoop	A tool designed for transferring bulk data between Hadoop and structured data stores such as relational database

According to IDC, the global market size of Hadoop projects in 2011 was US$77 million. The market grow almost ninefold to US$682.5 million by 2015 as shown in Figure 3-3.

4. Opportunities

Semi-structured and unstructured data sets are the two fastest growing data types in the digital universe. Analysis of these two data types will not be possible with traditional Database Management Systems. Hadoop HDFS and MapReduce enable the analysis of these two data types, giving organizations the opportunity to extract insights from bigger data sets within a reasonable amount of processing time.

Hadoop MapReduce's parallel processing capability has increased the speed of extraction and transformation of data. Hadoop MapReduce can be used as a data integration tool by reducing large amount of data to its representative form which can then be stored in the data warehouse.

At the current stage of development, Hadoop is not meant to be a replacement for scale-up storage and is designed more for batch processing rather than for interactive applications. It is also not optimized to work on small file sizes as the performance gains may not be considered when compared to huge data processing.

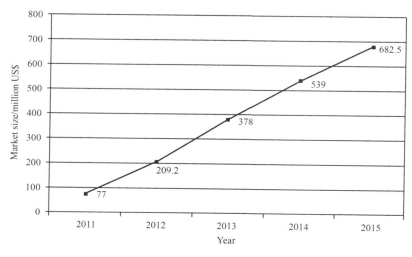

Figure 3-3 The global market size of Hadoop project.

5. Inhibitors

Lack of industry standards is a major inhibitor to Hadoop adoption. Currently, a number of emerging Hadoop vendors are offering their customized versions of Hadoop. HDFS is not fully Portable Operating System Interface (POSIX) compliant, which means system administrators cannot interact with it the same way they would with a Linux or UNIX system.

The scarcity of expertise capable of building a MapReduce system, managing Hadoop applications and performing deep analysis of the data will also be a challenge. At the same time, many of the Hadoop projects require customization and there is no industry standard on the best practices for implementation.

Many enterprises may not need Hadoop on a continual basis except for specific batch processing jobs that involve very huge data sets. Hence, the Return on Investment (ROI) for on-premise deployment will not be attractive. In addition, the integration of Hadoop clusters with legacy systems is complex and cumbersome, and is likely to remain a big challenge in the near term.

3.3.3 NoSQL Database Management System (NoSQL DBMS)

NoSQL (Not Only Structured Query Language) as a prime representative of the cloud data management system is widely applied. Many large IT systems adopt NoSQL as the main approach to data management. NoSQL as the substitute for a traditional database is the system collection with similar features rather than some data management system or a database. Thus, a lot of people consider NoSQL as an ecosystem instead of a system.

1. Three Popular Types of NoSQL Databases

1) Key-value store

As the name implies, key-value store is a system that stores values indexed for retrieval by keys. This system can hold structured or unstructured data. Amazon's SimpleDB is a Web service that provides core database functions of information indexing and querying in the cloud. It provides a simple API for storage and access. Users only pay for the services they use.

2) Columnar store

Rather than stores information in a heavily structured table of columns and rows with uniform-sized fields for each record, as is the case with relational database, columnar database contains one extendable column of closely related data. For example, Facebook create the high performance Cassandra to help power its website; The Apache Software Foundation developed Hbase, a distributed open source database that emulates Google's BigTable.

3) Document store

This database stores and organizes data as collections of documents, rather than as structured tables with uniform-sized fields for each record. With this database, users can add any number of fields of any length to a document. For example, The Apache Software Foundation hosts CouchDB as an open source, scalable database written in Erlang and accessible from any browser; Basho Technologies' Riak is a distributed, scalable, decentralized, open source database suitable for Web-based applications.

2. Advantages

NoSQL database generally process data faster than relational database. Relational database usually subject all data to the same set of ACID (atomicity, consistency, isolation, durability) restraints. Atomicity means an update is performed completely or not at all, and consistency means no part of a transaction will be allowed to break a database's rules. Isolation means each application operates concurrently, and durability means that completed transactions will persist. Having to perform these restraints on every piece of data makes relational database slower. Developers usually don't have their NoSQL database support ACID in order to increase performance, but this can cause problems when used for applications that require great precision. NoSQL database operates faster because its data models are simple. Thus, there is a bit of trade-off between speed and model complexity. Because NoSQL database doesn't have all the technical requirements that relational database does, most major NoSQL systems are flexible enough to better enable developers to use the applications in ways that meet their needs.

3. Problems

NoSQL database face several challenges.

1）Overhead and complexity

Because NoSQL database don't work with SQL, they require manual query programming, which can be fast for simple tasks but time-consuming for others. In addition, complex query programming for the databases can be difficult.

2）Reliability

Relational database natively supports ACID, while NoSQL database doesn't. NoSQL database thus doesn't natively offer the degree of reliability that ACID provides. If users want NoSQL database to apply ACID restraints to a data set, they must perform additional programming.

3）Consistency

Because NoSQL database doesn't natively support ACID, it could compromises consistency that enables better performance and scalability but is a problem for certain types of applications and transactions, such as those involved in banking.

4）Unfamiliarity with the technology

Most organizations are unfamiliar with NoSQL database and thus may not feel knowledgeable enough to choose one or even to determine that the approach might be better for their purposes.

5）Limited ecostructure

Unlike commercial relational database, many open source NoSQL applications don't yet come with customer support or management tools.

3.4 New Words & Phrases & Sentences

3.4.1 New Words

1. stack — *n.* 一叠，一摞，一堆；大量
   v. （使）放成整齐的一叠（或一摞、一堆）
2. synonymous — *adj.* 同义的，等同于
3. infrastructure — *n.* （国家或机构的）基础设施，基础建设
4. multi-GigaBytes — *n.* 数 GB
5. exclusive — *n.* 独家新闻，独家报道
   adj. 专用的，独有的；排斥的，排他的
6. standalone — *n.* 脱机
   adj. 单独的，独立的
7. eliminate — *adj.* 排除；清除；消除；淘汰；消灭，干掉（尤指敌人或对手）
8. optimization — *n.* 最优化；优化；最佳化；优化系统
9. scalability — *n.* 可扩展性；可伸缩性；可量测性
10. exponential — *n.* 指数的，幂数的
11. redundancy — *n.* 冗余
12. capability — *n.* （实际）能力，性能，容量
13. query — *n.* 疑问；询问；问号；查询
    v. 怀疑；表示疑虑；询问

14. exhaustive　　　　　　　　　　*adj.* 详尽的；彻底的；全面的
15. scarcity　　　　　　　　　　　*n.* 缺乏；不足；稀少
16. representative　　　　　　　　*n.* 代表

　　　　　　　　　　　　　　　　adj. 典型的，有代表性的
17. paradigm　　　　　　　　　　　*n.* 典范；范例；样式；词形变化表
18. embed　　　　　　　　　　　　*v.* 把……牢牢地嵌入（插入、埋入）；

　　　　　　　　　　　　　　　　派遣（战地记者、摄影记者等）；嵌入
19. consistency　　　　　　　　　　*n.* 一致性；相容性
20. scalable　　　　　　　　　　　*adj.* 可升级的

3.4.2　Phrases

1. skewed conclusion　　　　　　　　　　　错误的结论
2. document-based store　　　　　　　　　基于文档的存储
3. limited ecostructure　　　　　　　　　有限的生态结构
4. Hadoop Distributed File System (HDFS)　Hadoop 分布式文件系统
5. transactional decision management　　　交易决策管理
6. in addition to　　　　　　　　　　　　除了
7. Database Management System　　　　　　数据库管理系统
8. parallel processing capability　　　　并行处理能力
9. huge data processing　　　　　　　　　海量数据处理
10. project-based decision management software　基于项目的决策管理软件
11. fraud detection　　　　　　　　　　　欺诈检测
12. airline pricing optimization　　　　　航空公司定价优化
13. lack of　　　　　　　　　　　　　　　缺乏
14. local computation and storage　　　　本地计算和存储

15. hardware failure 硬件故障
16. Redundant Array of Independent Disks (RAID) 独立磁盘冗余阵列
17. be familiar with 熟悉，精通；同……相好
18. business analytics 商业分析

3.4.3 Sentences

1. There are no comprehensive Big Data technology standards in place today. The main reason is that the Big Data analytics projects companies taking on are typically complex and diverse in nature. 目前还没有全面的大数据技术标准。主要原因是企业实施的大数据分析项目在本质上通常是复杂和多样的。

2. The Hadoop Distributed File System enables a highly scalable, redundant data storage and processing environment that can be used to execute different types of large-scale computing projects. Hadoop 分布式文件系统提供了一个高度可扩展、冗余的数据存储和处理环境，可用于执行不同类型的大型计算项目。

3. Storage systems are also becoming more flexible and are being designed in a scale-out fashion, enabling the scaling of system performance and capacity. 存储系统也变得越来越灵活，并以扩展的方式设计，实现了系统性能和容量的扩展。

4. In-memory computing is supported by increased capabilities in system memory delivered at lower prices, making multi-GigaBytes (even multi-TeraBytes) memory more affordable. 内存计算由价格较低的系统内存中的增强功能提供支持，这使得数 GB（甚至数 TB）的内存更加便宜。

5. Two architectures — the extended Relational Database Management System (RDBMS) and the NoSQL Database Management System — have been developed to manage the different types of data. 为了管理不同类型的数据，开发了两种体系架构——扩展关系型数据库管理系统（RDBMS）和 NoSQL 数据库管理系统。

6. There are tools that allow for highly complex and predictive analysis as well as tools that simply help with basic data aggregation and trend reporting. 有些工具可以进行高度复杂且具有预测性的分析，还有些工具仅有助于基本数据聚合和趋势报告。

7. In any case, the usage of the tools is not mutually exclusive — there can be a set of tools with different features residing in a system to enable Big Data analytics. 在任何情况下，工具的使用

都不是互斥的——系统中可能存在一组具有不同功能的工具，以支持大数据分析。

8. The process of data analysis usually involves a closed-loop decision making model, which at the minimum, includes steps such as track, analyse, decide and act. 数据分析过程通常涉及一个闭环决策模型，该模型至少包括跟踪、分析、决策和行动等步骤。

9. From a technology perspective, additional functionalities such as decision capture and retention are required to support collaboration and risk management. 从技术角度来看，需要额外的功能（如决策捕获和保留）来支持协作和风险管理。

10. Hadoop is a framework that provides open source libraries for distributed computing using MapReduce software and its own Distributed File System, simply known as the Hadoop Distributed File System (HDFS). Hadoop 是一个框架，它使用 MapReduce 软件和它自己的分布式文件系统（Hadoop 分布式文件系统，HDFS）为分布式计算提供开源库。

11. It is designed to scale-out from a few computing nodes to thousands of machines, each offering local computation and storage. 它被设计为从几个计算节点扩展到数千台机器，为每台机器提供本地计算和存储。

12. HDFS is a fault-tolerant storage system that can store huge amount of information, scales up incrementally and survives storage failure without losing data. HDFS 是一个容错存储系统，它可以存储大量信息，逐步扩展并在不丢失数据的情况下避免存储故障。

13. HDFS manages storage on the cluster by breaking files into small blocks and storing duplicated copies of them across the pool of nodes. HDFS 通过将文件分成小块并在节点池中存储它们的副本来管理集群上的存储。

14. Most enterprise data management tools (Database Management Systems) are designed to make simple queries run quickly. 大多数企业的数据管理工具（数据库管理系统）都是为了使简单的查询快速运行而设计的。

15. In addition to MapReduce and HDFS, Hadoop also refers to a collection of other software projects that uses the MapReduce and HDFS framework. 除了 MapReduce 和 HDFS，Hadoop 还引用了许多使用 MapReduce 和 HDFS 框架的其他软件项目。

16. At the current stage of development, Hadoop is not meant to be a replacement for scale-up storage and is designed more for batch processing rather than for interactive applications. 在目前的开发阶段，Hadoop 本不是纵向扩展存储的替代品，它更多被用于批处理，而不是用于交互式应用。

17. The scarcity of expertise capable of building a MapReduce system, managing Hadoop applications and performing deep analysis of the data will also be a challenge. 缺乏能够构建 MapReduce 系统、管理 Hadoop 应用和对数据进行深入分析的专业知识也将是一项挑战。

18. Rather than stores information in a heavily structured table of columns and rows with uniform-sized fields for each record, as is the case with relational database, columnar database contains one extendable column of closely related data. 关系型数据库将信息存储在一个由列和行组成的高度结构化的表中，该表为每项记录提供统一大小的字段，与关系型数据库不同，列式数据库包含一列可扩展的密切相关的数据。

19. This database stores and organizes data as collections of documents, rather than as structured tables with uniform-sized fields for each record. 文档数据库存储数据并将其组织为文档的集合，而不是为每项记录提供字段大小统一的结构化表。

3.5 Exercises

【Ex. 1】Content Questions.

1. Which two components constitute Big Data technology stack?

2. What is the infrastructure of Big Data?

3. Which two architectures constitute Big Data?

4. What is "two decision support and automation software categories"?

5. What is HDFS?

6. What is the role of Hadoop MapReduce's parallel processing capability?

7. What is the major inhibitor to Hadoop adoption?

【Ex. 2】句子翻译。

1. Using this method, each developer can provide their own physical path definition to this variable.

2. All the data is dumped into the main computer.

3. All of the configuration and code is already implemented in the sample.

4. This is about the simplest weightless thread scheduler you could choose.

5. Such models align with agile thinking.

6. This can result in a variety of scalability and maintenance problems.

7. This allows the storage nodes to replicate data when a device is found to have failed.

8. Scalable bandwidth provides the solution while offering more efficient use of network resources.

9. Redundancy and dependability give the cloud another edge.

10. The most fundamental reason for a software company to localize products is to increase total revenue and net income.

【Ex. 3】将下列词填入适当的位置（每词只用一次）。

| bottom | duplicate | machines | special | node |
| source | collected | operations | completion | individual |

1）Background of Hadoop

With an increase in the penetration of the Internet and the usage of the Internet, the data captured by Google increased exponentially year on year. Just to give you an estimate of this number, in 2007 Google (1) on an average 270 PB of data every month. The same number increased to 20000 PB everyday in 2009. Obviously, Google needed a better platform to process such enormous data. Google implemented a programming model called MapReduce, which could process this 20000 PB per day. Google ran these MapReduce operations on a (2) file system called Google File System (GFS). Sadly, GFS is not an open source.

Doug cutting and Yahoo! reverse engineered the model GFS and built a parallel Hadoop Distributed File System (HDFS). Software or framework that supports HDFS and MapReduce is known as Hadoop. Hadoop is an open (3) and distributed by Apache.

2）The framework of Hadoop Processing

Let's draw an analogy from our daily life to understand the working of Hadoop. The bottom of the pyramid of any firm is the people who are (4) contributors. They can be analysts, programmers, manual labors, chefs, etc. Managing their work is the project manager. The project manager is responsible for a successful (5) of the task. He needs to distribute labor and smoothen the coordination among them. Also, most of these firms have a people manager, who is more concerned about retaining the headcount.

Hadoop works in a similar format. On the ___(6)___ we have machines arranged in parallel. These machines are analogous to individual contributors in our analogy. Every machine has a data node and a task tracker. Data node is also known as HDFS (Hadoop Distributed File System) and task tracker is also known as MapReducers.

Data node contains the entire set of data and task tracker does all the ___(7)___. You can imagine task tracker as your arms and legs, which enables you to do a task and data node as your brain, which contains all the information that you want to process. These ___(8)___ are working in silos and it is very essential to coordinate them. The task trackers (project manager in our analogy) in different machines are coordinated by a job tracker. The job tracker makes sure that each operation is completed and if there is a process failure at any ___(9)___, it needs to assign a duplicate task to some task tracker. The job tracker also distributes the entire task to all the machines.

A name node, on the other hand, coordinates all the data nodes. It governs the distribution of data going to each machine. It also checks for any kind of purging which has happened on any machine. If such purging happens, it finds the ___(10)___ data which was sent to other data node and duplicates it again. You can think of this name node as the people manager in our analogy which is concerned more about the retention of the entire data set.

3.6 Translations for Dialogue

随着大数据越来越受欢迎，汤姆想多了解一些与大数据有关的知识。

汤姆：大数据是一种挑战吗？

迈克：的确如此！企业刚开始接触大数据，它是一个相对较新的现象。挑战是要在数据到达时对其进行组织，在必要时对其进行处理，并将其存储在易于检索的位置。

汤姆：听起来很有趣。

迈克：是的。大数据通常由多种类型的数据组成，如文本、图像、视频、数字和音频，这些类型的数据是由交易处理系统、传感器、社交媒体、智能手机和其他多种多样的数据源生成的。

珍妮：然而，有些数据可能是结构化的，而其他数据是非结构化的。组织这种混合数据是一项挑战，可能需要数据库设计者寻求关系型数据库的严格结构以外的解决方案。

汤姆：你说的对。顺便问一下，低密度数据是怎么回事？

迈克：低密度数据是指大量非常详细的数据，其中的许多细节并不重要。高密度数据与之相反，它包含了大量有用的信息。

汤姆：我明白了。谢谢你的指点。

3.7 Translations for Reading

3.7.1 大数据技术栈

目前还没有全面的大数据技术标准。主要原因是企业实施的大数据分析项目在本质上通常是复杂和多样的。

Hadoop 在业界几乎是"大数据"的同义词，在处理大量非结构化数据方面很受欢迎。Hadoop 分布式文件系统提供了一个高度可扩展、冗余的数据存储和处理环境，可用于执行不同类型的大型计算项目。

大数据技术栈可以分为两个主要成分：硬件成分和软件成分，如图 3-1 所示。硬件成分指的是基础设施。软件成分可以进一步分为数据组织和管理软件、数据分析和发现软件、决策支持和自动化软件。

1. 基础设施

基础设施是大数据技术栈的基础。任何数据存储基础设施的主要成分（工业标准 x86 服务器和 10 Gbps 的网络带宽）都可以扩展为大型数据存储设施。存储系统也变得越来越灵活，并以扩展的方式设计，实现了系统性能和容量的扩展。内存计算由价格较低的系统内存中的增强功能提供支持，这使得数 GB（甚至数 TB）的内存更加便宜。

2. 数据组织和管理软件

这一层是指处理和准备所有类型的结构化和非结构化数据以供分析的软件。该层提取、清洗、规范和集成数据。为了管理不同类型的数据，开发了两种体系架构——扩展关系型数据库管理系统（RDBMS）和 NoSQL 数据库管理系统。

采用一些方法优化了扩展 RDBMS 处理大型关系数据（即结构化数据）集的规模和速度，

如使用列式存储来减少表扫描（列式数据库）数量和利用大规模并行处理（MPP）架构等。另外，NoSQL 数据库管理系统（NoSQL DBMS）产生于认识到 SQL 的交易质量和详细索引不适用于非结构化文件的处理。

3. 数据分析和发现软件

这一层包括两个数据分析软件部分——支持离线、即席、发现和深度分析的软件，以及支持动态实时分析和基于规则的自动化交易决策软件。这些工具还可以按分析数据的类型进行分类，如文本、音频和视频。这一层中的工具的复杂程度不同。有些工具可以进行高度复杂且具有预测性的分析，还有些工具仅有助于基本数据聚合和趋势报告。在任何情况下，工具的使用都不是互斥的——系统中可能存在一组具有不同功能的工具，以支持大数据分析。

4. 决策支持和自动化软件

数据分析过程通常涉及一个闭环决策模型，该模型至少包括跟踪、分析、决策和行动等步骤。在数据分析的基础上，支持决策并确保采取行动并非易事。从技术角度来看，需要额外的功能（如决策捕获和保留）来支持协作和风险管理。

有两类决策支持和自动化软件：交易决策管理软件和基于项目的决策管理软件。前者是自动化的，嵌入在应用程序中，本质上是实时且基于规则的。它允许使用输出来规定或执行规则、方法和进程。例如，欺诈检测、证券交易、航空公司定价优化、产品推荐和网络监控。基于项目的决策管理软件本质上通常是独立的、临时的和具有探索性的。它可以用来预测和估计趋势，示例包括定向市场营销、产品开发和天气预报的客户细分应用。

3.7.2 Hadoop MapReduce 和 Hadoop 分布式文件系统（HDFS）

Hadoop 是一个框架，它使用 MapReduce 软件和它自己的分布式文件系统（Hadoop 分布式文件系统，HDFS）为分布式计算提供开源库。它被设计为从几个计算节点扩展到数千台机器，为每台机器提供本地计算和存储。Hadoop 的主要价值主张之一是它被设计为在商品服务器或个人计算机等商品硬件上运行，并且对硬件故障有很高的容忍度。在 Hadoop 中，硬件故障被视为规则而不是异常。

1. HDFS

HDFS 是一个容错存储系统，它可以存储大量信息，逐步扩展并在不丢失数据的情况下避免存储故障。Hadoop 集群是用价格较低的计算机构建的。如果一台计算机（或节点）出现故障，只需将工作重新分配给集群中的其他计算机，集群即可继续运行，不会丢失数据或中断工作。HDFS 通过将文件分成小块并在节点池中存储它们的副本来管理集群上的存储。使用 HDFS 进行分布式文件存储的示例如图 3-2 所示。在本例中，即使有两台服务器发生故障，整个数据集仍然可用。

与其他冗余技术（包括独立磁盘冗余阵列机器所采用的策略）相比，HDFS 具有两个关键优势。首先，HDFS 不需要特殊的硬件，因为它可以从普通硬件中构建；其次，它以 MapReduce 的形式实现了一种高效的数据处理技术。

2. MapReduce

大多数企业的数据管理工具（数据库管理系统）都是为了使简单的查询快速运行而设计的。通常对数据建立索引，这样只需检查数据的一小部分就可以给出查询结果。但是，该方法不适用于无法索引的数据，即半结构化数据（文本文件）或非结构化数据（媒体文件）。在这种情况下，要给出查询结果，就必须检查所有数据。Hadoop 使用 MapReduce 技术快速进行详尽的分析。MapReduce 是一种使用并行编程实现的数据处理算法。简单来说，MapReduce 是一种编程范例，它涉及在运行"map"函数的多个节点上分配任务。map 函数获取问题，将其拆分为子问题并发送给不同的机器，以便所有的子问题可以同时运行。并行映射函数的结果被收集并分发到一组运行"reduce"函数的服务器上，然后这些服务器获取子问题的运行结果，并将其重新组合，以得到唯一解。

3. Hadoop 生态系统

除了 MapReduce 和 HDFS，Hadoop 还引用了许多使用 MapReduce 和 HDFS 框架的其他软件项目，表 3-1 简要介绍了其中的一些工具。

IDC 的数据显示，2011 年 Hadoop 项目的全球市场规模为 7700 万美元。2015 年，该市场增长了近 9 倍，达到 6.825 亿美元，如图 3-3 所示。

4. 机会

半结构化和非结构化数据集是数字世界中增长最快的两种数据。传统的数据库管理系统无法分析这两种数据。Hadoop HDFS 和 MapReduce 支持对这两种数据的分析，使企业有机会在合理的处理时间内从更大的数据集中获取见解。

Hadoop MapReduce 的并行处理能力提高了数据的提取和转换速度。Hadoop MapReduce 可以作为数据集成工具，将大量的数据缩减为其代表形式，然后将其存储在数据仓库中。

在目前的开发阶段，Hadoop 本不是纵向扩展存储的替代品，它更多被用于批处理，而不是用于交互式应用。它也没有针对小文件大小进行优化，因为与海量数据处理相比，可能不会考虑性能提高。

5. 障碍

缺乏行业标准是 Hadoop 应用的主要障碍。目前，许多新兴的 Hadoop 供应商都提供了他们定制的 Hadoop 版本。HDFS 不完全兼容可移植操作系统接口（POSIX），这意味着系统管理员不能像对待 Linux 或 UNIX 系统那样与 HDFS 进行交互。

缺乏能够构建 MapReduce 系统、管理 Hadoop 应用和对数据进行深入分析的专业知识也将是一项挑战。同时，许多 Hadoop 项目都需要定制，并且没有实施最佳实践的行业标准。

除了涉及大量数据集的特定批处理工作，许多企业可能不需要持续使用 Hadoop。因此，用于本地部署的投资回报率（ROI）不再有吸引力。此外，Hadoop 集群与遗留系统的集成复杂且烦琐，在短期内可能仍然是一项巨大的挑战。

3.7.3　NoSQL 数据库管理系统（NoSQL DBMS）

作为云数据管理系统的主要代表，NoSQL 得到了广泛应用。许多大型 IT 系统都将 NoSQL 作为数据管理的主要方式。NoSQL 取代了传统的数据库，它是具有相似功能的系统集合，而不是一些数据管理系统或一个数据库。因此，很多人认为 NoSQL 是一个生态体系而不只是一个系统。

1. 3 种受欢迎的 NoSQL 数据库管理系统

1）键值存储

顾名思义，键值存储是一个存储用于键值检索的索引值的系统。该系统可以保存结构化或非结构化数据。亚马逊的 SimleDB 是一种网络服务，它提供云信息索引和核心数据库查询功能，为存储和访问提供了一个简单的应用程序编程接口。用户只为他们使用的服务付费。

2）列式存储

关系数据库将信息存储在一个由列和行组成的高度结构化的表中，该表为每项记录提供统一大小的字段，与关系型数据库不同，列式数据库包含一列可扩展的密切相关的数据。例如，Facebook 创建高性能的 Cassandra 来帮助其网站发展；Apache 软件基金会开发了 Hbase（一个模拟 Google BigTable 的分布式开源数据库）。

3）文档存储

文档数据库存储数据并将其组织为文档的集合，而不是为每项记录提供字段大小统一的结构化表。用户可以使用这些数据库在文档中添加任意长度、任意数量的字段。例如，Apache 软件基金会将 CouchDB 作为一个用 Erlang 编写的、可以用任何浏览器访问的可扩展的开源数据库；Basho Technologies 的 Riak 是一个适用于基于网络的应用的分布式、可伸缩、分散的开源数据库。

2. 优点

NoSQL 数据库处理数据的速度通常比关系型数据库快。关系型数据库通常将所有数据置于同一组 ACID（原子性、一致性、隔离性、持久性）约束下。原子性表示所有操作要么全部执行，要么全部不执行；一致性表示事务的任何部分都不允许破坏数据库的规则，隔离性表示每个应用同时运行，持久性表示已完成的事务将持续存在。必须对每个数据执行这些限制会使关系型数据库的速度变慢。为了提高性能，开发人员通常不让 NoSQL 数据库支持 ACID，但是当用于具有较高精确性要求的应用时，这可能会导致出现问题。NoSQL 数据库运行更快，因为其数据模型很简单。因此，需要在速度和模型复杂度之间进行权衡。由于 NoSQL 数据库没有关系型数据库所拥有的所有技术要求，所以大多数主要的 NoSQL 系统都足够灵活，以便使开发人员以满足其需求的方式更好地使用应用程序。

3. 问题

NoSQL 数据库面临一些挑战。

1）开销与复杂性

因为 NoSQL 数据库不能与 SQL 一起工作，所以它们需要手动查询编程，这对于简单的任务来说很快，但对于其他任务来说则非常耗时。此外，对数据库进行复杂的查询编程可能很困难。

2）可靠性

关系型数据库本身支持 ACID，而 NoSQL 数据库则不支持。因此，NoSQL 数据库本身不具有 ACID 提供的可靠性。如果用户希望 NoSQL 数据库对数据集应用 ACID 限制，则必须运行其他程序。

3）一致性

由于 NoSQL 数据库本身不支持 ACID，所以它可能会损害一致性，从而提高性能和可扩展性。但对于某些类型的应用和事务（如涉及银行业务的应用和事务）来说，这是一个难题。

4）对技术不熟悉

大多数企业都不熟悉 NoSQL 数据库，因此可能没有足够的知识来选择数据库，甚至没有足够的知识来确定这种方法可能更适合他们的用途。

5）有限的生态结构

与商用关系型数据库不同，许多开源 NoSQL 应用尚未附带客户支持或管理工具。

Unit Four Big Data Target Analytics

4.1 Learning Goals

After learning this unit, you will be able to master the following knowledge:

1. Text analytics is the process of deriving information from text sources.

2. In-memory analytics is an analytics layer in which detailed data (up to TeraByte size) is loaded directly into the system memory from a variety of data sources, for fast query and performance calculation.

3. Predictive analytics is a set of statistical and analytical techniques that are used to uncover relationships and patterns within a large volume of data that can be used to predict behaviour or events.

4. Software-as-a-Service (SaaS) is software owned, delivered and managed remotely by one or more providers.

5. Graph analytics is the study and analysis of data that can be transformed into a graph representation consisting of nodes and links.

6. A "complex" event is an abstraction of other "base" events and represents the collective significance of these events.

7. Mobile business analytics is an emerging trend that rides on the growing popularity of mobile computing.

8. Video analytics is the automatic analysis of digital video images to extract meaningful and relevant information.

9. Audio analytics uses a technique commonly referred to as audio mining where a large volume of audio data are searched for specific audio characteristics.

4.2 Dialogue

After class, Sophie and Henry are standing by the door, waiting for Mark.

Henry: Excuse me, Sophie. May I ask you some questions about Big Data?

Sophie: Sure. What can I do for you?

Henry: What does Big Data analytics mean?

Sophie: Big Data analytics is the process of examining large data sets to uncover hidden patterns, unknown correlations, market trends, customer preferences and other useful business information.

Henry: So interesting. Is Big Data analytics important?

Sophie: Of course. Big Data analytics helps organizations harness their data and use it to identify new opportunities. That, in turn, leads to smarter business moves, more efficient operations, higher profits and happier customers.

Henry: As far as I'm concerned, we can benefit a lot from Big Data analytics, such as cost reduction, faster and better decision making, provide new products and services.

Sophie: You said it! By the way, semi-structured data and unstructured data are different from traditional data based on relational database. As a result, many organizations looking to collect, process, and analyse Big Data have turned to a newer class of technologies that includes Hadoop and related tools such as Yarn, MapReduce, Spark.

Henry: Well, from your descriptions, I'm sure Big Data analytics will have a good future.

Sophie: I think so.

4.3 Reading

4.3.1 Basic Analytics

1. Text Analytics

Text analytics is the process of deriving information from text sources. These text sources are forms of semi-structured data that include Web-based materials, blogs and social media postings (such as tweets). The technology within text analytics comes from fundamental fields including linguistics, statistics and Machine Learning. In general, modern text analytics uses statistical models, coupled with linguistic theories, to capture patterns in human languages so that machines can "understand" the meaning of texts and perform various text analytics tasks. These tasks can be as simple as entity extraction or more complex in the form of fact extraction and concept extraction.

Entity extraction: entity extraction identifies an item, a person or any individual piece of information such as dates, companies or countries.

Fact extraction: a fact is a statement about something that exists, has happened and is generally known. Fact extraction identifies a role, relationship, cause or property.

Concept extraction: concept extraction identifies an event, process, trend or behaviour.

Text analytics will be an increasingly important tool for organizations as the attention shifts from structured data analysis to semi-structured data analysis. One major application of text analytics would be in the field of sentiment analysis where consumer feedback can be extracted from the social media feeds and blog commentaries. The potential of text analytics in this application has spurred much research interest in the R&D community.

1) Advantages

Combining text analytics with traditional structured data allows a more complete view of the issue, compared to analysis using traditional data mining tools. Applying text mining in the area of sentiment analysis helps organizations uncover sentiments to improve their Customer Relationship Management (CRM). Text analytics can also be applied in the area of public security by the scrutiny of text for patterns that characterise criminal activity or terrorist conspiracy.

2) Problems

The text analytics solution market is still immature. The analytics engine will face challenges in dealing with non-English content and local colloquialisms. Hence, a text analytics solution developed for a particular market may not be directly applicable in another situation (a certain level of customization will be needed). The suitability of text analytics on certain text sources, such as technical documents or documents with many domain specific terms, may be questionable as well.

Adoption of text analytics is more than just deploying the technology. Knowing the metrics to use in determining the results is also a required skill. There has to be an understanding of what to analyse and how to use the outcome of analysis to improve business. This requires a certain level of subjectivity which may not be what management desires.

2. In-memory Analytics

In-memory analytics is an analytics layer in which detailed data (up to TeraByte size) is loaded directly into the system memory from a variety of data sources, for fast query and performance calculation. In theory, this approach partly removes the need to build metadata in the form of relational aggregates and pre calculated cubes.

The use of in-memory processing as a backend resource for business analytics improves performance. On the traditional disk-based analytics platform, metadata has to be created before the actual analytics process takes place. The way which the metadata modeled is dependent on the analytics requirements. Changing the way to model the metadata to fulfill new requirements requires a good level of technical knowledge. In-memory analytics removes the need to pre-model this metadata for every end user's needs. The relevance of the analytics content is also improved as data can be analysed the moment it is stored in the memory. The speed that is delivered by in-memory analytics makes it possible to power interactive visualization of data sets, making data access a more exploratory experience, as shown in Figure 4-1.

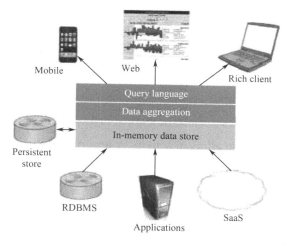

Figure 4-1　In-memory analytics overview.

In-memory analytics is enabled by a series of in-memory technologies. Taxonomy of in-memory analytics technologies is shown in Figure 4-2.

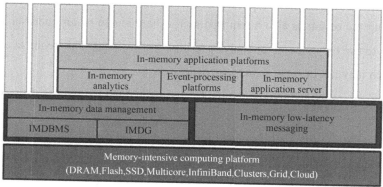

Figure 4-2　Taxonomy of in-memory analytics technologies.

In-memory data management:

In-Memory Database Management System (IMDBMS): an IMDBMS stores the entire database in the computer RAM, negating the need for disk I/O instructions. This allows applications to run completely in memory.

In-Memory Data Grid (IMDG): the IMDG provides a distributed in-memory store in which multiple, distributed applications can place and retrieve a large volume of data objects.

In-memory low-latency messaging:

This platform provides a mechanism for applications to exchange messages as rapidly as possible through direct memory communications.

1) Advantages

In-memory analytics can work with large and complex data sets and churn output in a matter of minutes or even seconds. This way, in-memory analytics can be provided as a form of real-time or near real-time services to users. With less dependency on the metadata, in-memory analytics will enable self-service analysis. When coupled with interactive visualization and data discovery tools for intuitive, unencumbered and fast exploration of data, in-memory analytics can facilitate a series of "what-if" analyses with quick results. This facilitates the fine-tuning of analytics results with multiple query iterations which, in turn, enables better decision making.

In-memory analytics can be adopted to complement other Big Data solutions. One possibility is to run in-memory analytics on a Hadoop platform to enable high-speed analytics on distributed data. NoSQL database could run on in-memory computing platform to support in-memory analytics. In-memory analytics can run on In-Memory Data Grid that can handle and process multi-PetaBytes of in-memory data sets.

2) Problems

In-memory analytics technology has the potential to subvert enterprise data integration efforts. As the technology enables standalone usage, organizations need to ensure they have the means to govern its usage and there is an unbroken chain of data lineage from the report to the original source system.

Although the average price per GigaByte of RAM has been going down, it is still much higher than the average price per GigaByte of disk storage. In the near term, the memory still exacts a premium relative to the same quantity of disk storage. Currently, Amazon Web Services charges about US$0.10 per GigaByte of disk storage while memory costs about 100 times as much, at about US$10.57 per GigaByte. The utilisation of scaled-up in-memory analytics is likely to be restricted to the deep-pocketed and technically astute organizations.

3. Predictive Analytics

Predictive analytics is a set of statistical and analytical techniques that are used to uncover relationships and patterns within a large volume of data that can be used to predict behaviours or events. Predictive analytics may mine information and patterns in structured and unstructured data sets as well as data streams to anticipate future outcomes. The real value of predictive analytics is to provide predictive support service that goes beyond traditional reactive break-and-fix assistance and toward a proactive support system by preventing service-impacting events from occurring.

According to Gartner, three methods are being evaluated within the marketplace for the prediction of technical issues within the product support space. Gartner believes that mature predictive support services will ultimately use a combination of all three of these approaches.

Pattern-based approach: these technologies seek to compare real-time system performance and configuration data with unstructured data sources that may include known failure profiles, historic failure records and customer configuration data. Powerful correlation engines are required to seek statistical patterns within huge, multifaceted repositories to determine if the customer's current configuration and performance data indicate a likely failure.

Rule-based approach: statistical analysis of historical performance data, previously identified failure modes and the results of stress or load testing is used to define a series of rules that real-time telemetry data is compared with. Each rule may interrogate multiple telemetry data points and other external factors such as the time of day, environmental conditions and concurrent external activities, against defined thresholds. Breaches of these rules may then be collated and escalation routines can be triggered, depending on the likely severity and impact of the resultant issues or outages.

Statistical process control-based approach: control chart theory has been a mainstay of quality manufacturing processes for more than half a century and has proven an invaluable aid to manage complex process-driven systems. The advent of retrofit capable, real-time telemetry, improvements in data acquisition solutions and network capacity to support such a large volume of data means the statistical techniques that underpinned the quality revolution within the manufacturing space can now be used to industrialise IT service delivery. Statistical anomalies can be readily identified and used to initiate appropriate contingent or preventive actions to ensure that service performance is unaffected and the business can continue to function as normal.

1）Advantages

Predictive analytics exploits patterns found in historical and transactional data to identify future risks and opportunities. Approaches are focused on helping companies acquire actionable insights, identify and respond to new opportunities more quickly.

Predictive analytics has direct applications in many verticals:

Law enforcement agencies can look for patterns in criminal behaviours and suspicious activities which can help them identify possible motives and suspects, leading to the more effective deployment of personnel. For example, the Edmonton Police Services use sensors and data analytics to create maps that define high-crime zones so that additional police resources can be proactively diverted to them.

Public health authorities can monitor information from various sources, looking for elevated levels of certain symptoms that signal a potential outbreak of disease. In Singapore, the Institute of High Performance Computing (IHPC) has developed a simulation model for epidemiological study. In the case of a pandemic, the model can be used to predict the spread and Tax departments can use predictive analytics to identify patterns to highlight cases that warrant further investigation. Additionally, predictive analytics can be used to understand the likely impact of policies on revenue generation.

2) Problems

As predictive analytics requires a multitude of data inputs, the challenge is to know which data inputs would be relevant and how these disparate sources can be integrated. In some predictive models, there is a need for user data touches on the sensitive issue of data privacy. Finally, the reliability of the prediction outcome faces scepticism, especially when the prediction outcome deviates from the decision maker's point of view. It will take time for decision makers to accept the outcomes of predictive analytics as influencing factors in decision. Similarly, it will take time for the underlying predictive algorithms to progress and mature.

4. SaaS-based Business Analytics

Software-as-a-Service (SaaS) is software owned, delivered and managed remotely by one or more providers. A single set of common code is provided in an application that can be used by many customers at any time. SaaS-based business analytics enables customers to deploy one or more prime components of business analytics quickly without significant IT involvements or the need to deploy and maintain an on-premise solution.

The prime components are as follows.

Analytic applications: support performance management with prepackaged functionality for specific solutions.

Business analytics platforms: provide the environment for development and integration, information delivery and analysis.

Information management infrastructures: provide data architecture and data integration infrastructure.

1) Advantages

Leveraging the benefits of Cloud Computing, SaaS-based business analytics offers a quick, low-cost and easy-to-deploy business analytics solution. This is especially the case for enterprises that do not have the expertise to set up an in-house analytics platform nor the intention to invest in internal business analytics resources. SaaS-based business analytics may be useful for mid and small enterprises that have yet to invest in any form of on-premise business analytics solutions.

2) Problems

There could be integration challenges for enterprises that want to export data to, and extract data from service provider for integration with the on-premise information infrastructure. As the data resides on the cloud, SaaS-based business analytics may not suitable for businesses that have to worry about data privacy and security issues.

5. Graph Analytics

Graph analytics is the study and analysis of data that can be transformed into a graph representation consisting of nodes and links. Graph analytics is good for solving problems that do not require the processing of all available data within a data set. A typical graph analytics problem requires the graph traversal technique. Graph traversal is a process of walking through the directly connected nodes. An example of a graph analytics problem is to find out how many ways two members of a Social Network are linked directly and indirectly.

Different forms of graph analytics exist as follows.

Single path analysis: the goal is to find a path through the graph, starting with a specific node. All the links and the corresponding vertices that can be reached immediately from the starting node are first evaluated. From the identified vertices, one is selected based on a certain set of criteria and the first hop is made. After that, the process continues. The result will be a path consisting of a number of vertices and edges.

Optimal path analysis: this analysis finds the best path between two vertices. The best path could be the shortest path, the cheapest path or the fastest path, depending on the properties of the vertices and the edges.

Vertex centrality analysis: this analysis identifies the centrality of a vertex based on several centrality assessment properties.

Degree centrality analysis: this analysis indicates how many edges a vertex has. The more edges there are, the higher the degree centrality.

Closeness centrality analysis: this analysis identifies the vertex that has the smallest number of hops to other vertices. The closeness centrality of the vertex refers to the proximity of the vertex in reference to other vertices. The higher the closeness centrality, the more number of vertices that require short paths to the other vertices.

Eigenvector centrality analysis: this analysis indicates the importance of a vertex in a graph. Scores are assigned to vertices, based on the principle that connections to high-scoring vertices contribute more to the score than equal connections to low-scoring vertices.

Graph analytics can be used in many areas as follows.

In the finance sector, graph analytics is useful for understanding the money transfer pathways. A money transfer between bank accounts may require several intermediate bank accounts and graph analytics can be applied to determine the different relationships between different account holders. Running the graph analytics algorithm on the huge financial transaction data sets will help to alert banks to possible cases of fraudulent transactions or money laundering.

The use of graph analytics in the logistics sector is not new. Optimal path analysis is the obvious form of graph analytics that can be used in logistics distribution and shipment environments. There are many examples of using graph analytics in this area and they include "the shortest route to deliver goods to various addresses" and the "most cost-effective routes for goods delivery".

One of the most contemporary use cases of graph analytics is in the area of social media. It can be used not just to identify relationships in Social Networks, but to understand them. One outcome from using graph analytics in the area of social media is to identify the "influential" figures from each social graph. Businesses can then spend more effort in engaging this specific group of people in their marketing campaigns or Customer Relationship Management (CRM) efforts.

4.3.2　Complicated Analytics

1. Mobile Business Analytics

Mobile business analytics is an emerging trend that rides on the growing popularity of mobile computing. From the workforce's perspective, being able to access the latest analytics insights from backend data stores creates a need for mobile business analytics. There are two types of mobile business analytics: passive mobile business analytics and active mobile business analytics.

Passive mobile business analytics revolves around the "push" factor. Event-based alerts or reports can be pushed to the mobile devices after being refreshed at the backend. Passive mobile business analytics may be a step ahead in providing accessibility convenience to the users but it is not enough to support the just-in-time analytical requirements users want. Active mobile business analytics enables users to interact with the business analytics systems on-the-fly. It can work as a combination of both "push" and "pull" techniques. An initial view of a report could comprise "push" technique and further analytical operations on the report to obtain additional information could comprise "pull" technique.

There are two approaches to develop business analytics applications for mobile devices: develop softwares for specific mobile operating systems such as iOS or Android, or develop browser-based versions of their business analytics applications. Developing a browser-based version of the business analytics application requires only a one-time development effort as the deployment can be made across devices. Furthermore, custom-developed mobile business analytics applications can provide full interactivity with the content on the device. In addition, this approach provides periodic caching of data that can be viewed offline.

1）Advantages

Mobile business analytics facilitates off-site decision making. More often than not, decision makers only need access to a few key analytics metrics churned out by the backend data store. Having access to these metrics on a mobile device can reduce decision bottlenecks, improve the efficiency of business processes and enable broader inputs into the decision. In addition, device-specific capabilities can increase the value of mobile business analytics. For instance, incorporating location awareness with a mobile business analytics query provides a location context to the query. The user can then obtain the query results that are relevant to his location.

2) Problems

Technical and security risk concerns will inhibit the uptake of mobile business analytics, especially in the case of mission-critical deployments. Sensitive corporate data can be at risk if a tablet or smartphone device is compromised. The growth of mobile business analytics is dependent on the success of an enterprise's BYOD (Bring Your Own Device) programme.

2. Video Analytics

Video analytics is the automatic analysis of digital video images to extract meaningful and relevant information, also known as Video Content Analysis (VCA). It uses computer vision algorithms and machine intelligence to interpret, learn and draw inferences from image sequences. Video analytics automates scene understanding which otherwise would have required human monitoring. It is not only able to detect motion, but also qualifies the motion as an object, understands the context around the object, and tracks the object through the scene. In theory, any behaviour that can be seen and accurately defined on a video image can be automatically identified and subsequently trigger an appropriate response.

There are two systems in video analytics architecture, namely edge-based system and central-based system.

1) Edge-based system

Edge-based system is shown in Figure 4-3. In this system, video analytics is performed on the raw image on the video source (i.e., the camera) before any compression is applied. This means the image has the maximum amount of information content, allowing the analytics to work more effectively. The processing of locally sourced data makes the system resilient to transmission or network failures. Alerts can be responded locally or stored for transmission once the network connection is resumed. Bandwidth usage can be controlled by reducing frame rates, lowering the resolution and increasing image compression when no events or alerts are in progress. Edge-based system must be implemented through an IP video camera or video encoder with sufficient processing power. This may result in larger units with greater power requirements compared to conventional analogue or network cameras.

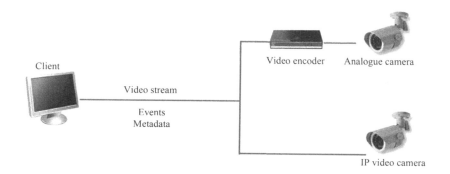

Figure 4-3 Edge-based system.

2) Central-based system

Central-based system is shown in Figure 4-4. In this system, video analytics is implemented through a dedicated server that pulls the video, analyses the video and issues the alerts or analysis results. Since the analytics equipment is installed in one central location, it makes for easier maintenance and upgrading. The system can be protected from power failure by using a central Uninterrupted Power Supply (UPS). While this independent-of-camera approach is applicable to most types of surveillance system, large amounts of network bandwidth may be required to transmit high-quality images from the image capture devices. Compression and transmission effects may impede the efficiency and accuracy of the analysis because of the loss of information contained within the images.

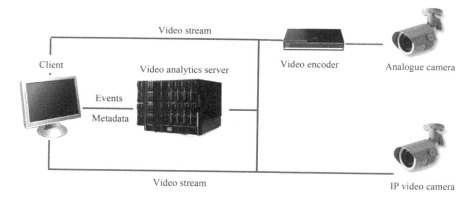

Figure 4-4 Central-based system.

3) Advantages

Video analytics can complement Business Intelligence with security awareness. Automated surveillance cameras with analytics can detect the presence of people and vehicles, and interpret their activities. Suspicious activities such as loitering or moving into an unauthorized area are automatically flagged and forwarded to security personnel. In the retail sector, video analytics can enhance the customer experience and drive revenue generation through an improved understanding of the customer experience. With video analytics, retailers are able to analyse customer time in the store to evaluate the effectiveness of key store promotional areas. Retailers will also be able to determine the demographics of store visitors and recognize the buying patterns by correlating sales receipts to customer demographics.

4) Problems

Despite technological advancements lead to improved video capabilities such as scene and object recognition, current analytics is still focused on intrusion detection and crowd management where events happen randomly and often appeal to very specific verticals that require customizations. In addition, video analytics may still be plagued by accuracy and reliability issues, including problems caused by shadows and areas of high contrast in light. The video analytics system will have the issue of false alert, largely due to the sophistication level of the analytics algorithms. False Alert Rate (FAR) refers to the number of times the system creates an alert when nothing of interest is happening. FAR of one per camera per day may sound acceptable but it may be magnified after multiplying across hundreds of cameras. For example, in a network of 100 cameras, that will equate to a false alert every 14 minutes.

Hardware limitation is a concern. While video analysis can take place at the edge of devices, the deployment cost or hardware requirement for each camera, with computing capability installed in each camera to run complex video analytics algorithms will be high. In addition, video analytics that takes place at the central server will require high bandwidth while the image feeds for video analysis may suffer from attenuation and loss of details. This may impede the accuracy of the video analysis. There is a rise of IP-based cameras but a majority of the current infrastructure continues to employ analogue-based cameras. The delineation of both technologies will also result in an increase in the cost of adoption of video analytics.

3. Audio Analytics

Audio analytics uses a technique commonly referred to as audio mining where a large volume of audio data are searched for specific audio characteristics. When applied in the area of speech recognition, audio analytics identifies spoken words in the audio and puts them in a search file. The two most common approaches to audio mining are Large-Vocabulary Continuous Speech Recognition and phonetic recognition.

Large-Vocabulary Continuous Speech Recognition: Large-Vocabulary Continuous Speech Recognition (LVCSR) converts speech to text and then uses a dictionary to understand what is being said. The dictionary typically contains up to several hundred thousand entries which include generic words as well as industry and company specific terms. Using the dictionary, the analytics engine processes the speech content of the audio to generate a searchable index file. The index file contains information about the words it understood in the audio data and can be quickly searched for keywords and phrases to bring out the relevant conversations that contain the search terms.

Phonetic recognition: phonetic recognition does not require any conversion from speech to text but instead works only with sounds. The analytics engine first analyses and identifies sounds in the audio content to create a phonetic-based index. It then uses a dictionary of several dozen phonemes to convert a search term to the correct phoneme string. Finally, the system looks for the search terms in the index.

One difference between the phonetic recognition and the LVCSR relates to which stage (indexing or searching) is the most computationally intensive. With phonetic recognition, the rate at which the audio content can be indexed is many times faster than with LVCSR. During the search stage, however, the computational burden is larger for phonetic recognition system than for LVCSR system, where the search pass is typically a simple operation. Phonetic recognition does not require the use of complex language models as it can be run effectively without knowledge of which words were previously recognized. In contrast, knowledge of which words were previously recognized is vital for achieving good recognition accuracy in LVCSR system. LVCSR system must, therefore, use sophisticated language models, leading to a much greater computation load at the indexing stage and resulting in significantly slower indexing speed.

An advantage of the phonetic recognition is that an open vocabulary is maintained, which means that searches for personal or company names can be performed without the need to reprocess the audio. In LVCSR system, any word that was not known by the system at the time the speech was indexed can never be found. For example, a previously unknown term called "consumerization" is now popular. This word is not already in the dictionary of words used by a LVCSR system and this means the analytics engine can never pick out this word in the audio file that was processed by the system. In order to find matches for this new word, the LVCSR system has to be updated with a new dictionary that contains the word "consumerization" and all the audio has to be pre-processed again. This is a time-consuming process. This problem does not occur with phonetic audio analytics system because it works at the level of phonemes, not words. As long as a phonetic pronunciation for the word can be generated at search time, it will be able to find matches for the word with no reprocessing of audio required.

1) Advantages

Audio analytics used in the form of speech analytics enables organizations with a call centre to obtain market intelligence. The call records that a call centre maintains typically represent thousands of hours of the "voice of the customer" which represents customer insights that the organizations will want to extract and mine for market intelligence. Speech analytics can be applied across all the call records for a company to gather data from the broadest sources possible. More importantly, the data is gathered from actual customer interactions rather than recollections of the interactions at a later date. As a result, the data is more accurately placed in the actual context.

Government agencies and regulated organizations have to handle large amounts of recorded conversations for security and intelligence purposes. Manual transcription and analysis are not only slow, they risk either late or missed intelligence. Audio analytics can be applied to real-time audio stream and users can move straight to the specific point in the conversation recording where the word or phrase of interest is used, improving the monitoring processes.

2) Problems

There are limitations to the accuracy and reliability of audio analytics. It is difficult to have 100% accuracy in the identification of audio streams. In the case of speech analytics, the system may not be able to handle accented words. Moreover, even if the analytics algorithms were sophisticated enough to identify words to a high level of accuracy, understanding the contextual meaning of the words would still be a challenge.

Note:

The text is adapted from: https://www.studymode.com/essays/Bigdata-39129403.html.

4.4 New Words & Phrases & Sentences

4.4.1 New Words

1. abstraction　　　　　　　　　　　　　　*n.* 抽象概念；抽象；出神；心神专注；提取；抽取；分离

2. linguistics   *n.* 语言学
3. statistical   *adj.* 统计数据的；统计学的
4. semi-structured   *n.* 半结构化
                                  adj. 半结构式；半结构化的；半结构性
5. evaluate   *v.* 估计；评价；评估
6. metadata   *n.* 元数据
7. mapping   *n.* 映射
8. sentiment   *n.* （基于情感的）观点，看法；情绪；
                                  伤感，柔情，哀伤
9. terrorist   *n.* 恐怖主义者；恐怖分子
10. utilisation   *n.* 利用，使用
11. conspiracy   *n.* 密谋策划；阴谋
12. immature   *adj.* （行为）不成熟的，不够老练的，幼稚的；
                                  未长成的；发育未全的
13. multiple   *adj.* 数量多的；多种多样的
                                  n. 倍数
14. subvert   *v.* 颠覆
15. algorithm   *n.* 算法；计算程序
16. spur   *n.* 马刺；靴刺；鞭策；激励；刺激；鼓舞；
                                  尖坡；支脉
                                  v. 鞭策；激励；刺激；鼓舞；促进，加速
17. customization   *n.* 定制
18. metric   *adj.* 米制的；公制的；用公制测量的
                                  n. 指标；度量；诗体；韵文；诗韵
19. subjectivity   *n.* 主观性，主观

20.	aggregate	*n.* 总数；合计；骨料，集料
		adj. 总数的；总计的；聚合的
		v. 总计；合计；聚集
21.	taxonomy	*n.* 分类学；分类法；分类
22.	iteration	*n.* 迭代；（计算机）新版软件
23.	intention	*n.* 目的；意图；意向；愈合
24.	premium	*n.* 保险费；额外费用；附加费
		adj. 高昂的；优质的
25.	correlation	*n.* 相互关系；相关；关联；相关性

4.4.2 Phrases

1.	data mining	数据挖掘
2.	statistical model	统计模型
3.	entity extraction	实体提取
4.	concept extraction	概念提取
5.	the traditional disk-based analytics platform	传统的基于磁盘的分析平台
6.	in-memory analytics	内存分析
7.	in-memory low-latency messaging	内存低延迟消息传递
8.	predictive analytics	预测分析
9.	traditional reactive break-and-fix assistance	传统的被动中断和修复辅助
10.	graph representation	图示
11.	fact extraction	事实提取
12.	a multitude of data inputs	大量的数据输入
13.	graph analytics	图分析
14.	local colloquialism	地方口语

15. optimal path analysis　　　　　　　　最优路径分析

4.4.3 Sentences

1. In general, modern text analytics uses statistical models, coupled with linguistic theories, to capture patterns in human languages so that machines can "understand" the meaning of texts and perform various text analytics tasks. 通常，现代文本分析使用统计模型和语言理论来捕捉人类语言的模式，以使机器"理解"文本的意义并执行各种文本分析任务。

2. One major application of text analytics would be in the field of sentiment analysis where consumer feedback can be extracted from the social media feeds and blog commentaries. 文本分析的主要应用之一是在情感分析领域，在该领域，可以从社交媒体信息和博客评论中提取消费者的反馈。

3. Combining text analytics with traditional structured data allows a more complete view of the issue, compared to analysis using traditional data mining tools. 与使用传统数据挖掘工具进行分析相比，将文本分析与传统结构化数据结合可以更全面地了解问题。

4. The suitability of text analytics on certain text sources, such as technical documents or documents with many domain specific terms, may be questionable as well. 文本分析在某些文本源（如技术文档或具有许多特定领域术语的文档）上的适用性也可能存在问题。

5. In-memory analytics is an analytics layer in which detailed data (up to TeraByte size) is loaded directly into the system memory from a variety of data sources, for fast query and performance calculation. 内存分析是一个分析层，在该层中，详细数据（大小高达太字节）从各种数据源直接加载到系统内存中，以实现快速查询和性能计算。

6. The speed that is delivered by in-memory analytics makes it possible to power interactive visualization of data sets, making data access a more exploratory experience. 内存分析提供的速度使数据集的交互式可视化成为可能，使数据访问成为一种更具探索性的体验。

7. When coupled with interactive visualization and data discovery tools for intuitive, unencumbered and fast exploration of data, in-memory analytics can facilitate a series of "what-if" analyses with quick results. 当与交互式可视化和数据发现工具结合使用以实现直观、无阻碍和快速的数据浏览时，内存分析可以促进一系列"假设"分析并快速获得结果。

8. One possibility is to run in-memory analytics on a Hadoop platform to enable high-speed analytics on distributed data. NoSQL database could run on in-memory computing platform to

support in-memory analytics. 一种可能是在 Hadoop 平台上运行内存分析，以对分布式数据进行高速分析。NoSQL 数据库可以在内存计算平台上运行，以支持内存分析。

9. Predictive analytics is a set of statistical and analytical techniques that are used to uncover relationships and patterns within a large volume of data that can be used to predict behaviours or events. 预测分析是一组统计和分析技术，用于发现大量数据的关系和模式，这些数据可用于预测行为或事件。

10. Graph analytics is the study and analysis of data that can be transformed into a graph representation consisting of nodes and links. Graph analytics is good for solving problems that do not require the processing of all available data within a data set. 图分析是对可以转化为由节点和链路组成的图示的数据的研究和分析。图分析非常适用于解决不需要处理数据集中所有可用数据的问题。

4.5　Exercises

【Ex. 1】**Content Questions.**

1. What is text analytics?
2. What is in-memory analytics?
3. What is predictive analytics?
4. Which three methods are being evaluated for the predictive analytics?

【Ex. 2】补充空白部分的单词。

1. In general, modern text analytics uses statistical models, _____ with linguistic theories, to capture patterns in human languages so that machines can "understand" the meaning of texts and perform various text _____ tasks.

2. Text analytics will be an increasingly important tool for organizations as the attention shifts from _____ data analysis to _____ data analysis.

3. The _____ of text analytics in this application has _____ much research interest in the R&D community.

4. Applying text mining in the area of _____ analysis helps organizations uncover

sentiments to _____ their Customer Relationship Management (CRM).

5. _____ of text analytics is more than just _____ the technology.

6. On the traditional disk-based analytics platform, _____ has to be created before the actual analytics process takes place.

7. Developing a browser-based version of the _____ analytics application requires only a one-time development effort as the deployment can be made across _____.

8. Technical and _____ risk concerns will _____ the uptake of mobile business analytics, especially in the case of mission-critical deployments.

4.6 Translations for Dialogue

下课后，索菲和亨利站在门口等马克。

亨利：打扰一下，索菲。我可以向你询问一些关于大数据的问题吗？

索菲：好的，我能为你做些什么？

亨利：大数据分析是什么意思？

索菲：大数据分析是检查大型数据集以发现隐藏模式、未知的相关性、市场趋势、客户偏好和其他有用的商业信息的过程。

亨利：真有趣，大数据分析重要吗？

索菲：当然。大数据分析可以帮助企业利用其数据并使用它来发现新的机会。这反过来又会使企业的商业行为更明智、运营更高效、利润更高、客户更愉悦。

亨利：我认为，大数据分析使我们受益良多，如降低成本、做出更快和更好的决策，以及提供新产品和服务。

索菲：你说的对！另外，半结构化和非结构化数据与基于关系型数据库的传统数据不同。因此，许多希望收集、处理和分析大数据的企业都转而采用一类新的技术，其中包括 Hadoop 和相关工具，如 Yarn、MapReduce 和 Spark。

亨利：嗯，根据你的描述，我相信大数据分析会有一个美好的未来。

索菲：我也这样认为。

4.7 Translations for Reading

4.7.1 基础分析

1. 文本分析

文本分析是从文本源导出信息的过程。这些文本源为半结构化数据形式,包括网络资源、博客和社交媒体发布的帖子(如推文)。文本分析中的技术来自语言学、统计学和机器学习等基础领域。通常,现代文本分析使用统计模型和语言理论来捕捉人类语言的模式,以使机器"理解"文本的意义并执行各种文本分析任务。这些任务可以像实体提取一样简单,也可以以更复杂的事实提取和概念提取形式存在。

实体提取:实体提取识别一个项目、一个人或任何单一信息,如日期、企业或国家。

事实提取:事实是对存在的、已经发生的和众所周知的事物的陈述。事实提取识别角色、关系、原因或属性。

概念提取:概念提取识别事件、过程、趋势或行为。

随着人们的注意力从结构化数据分析转向半结构化数据分析,文本分析将成为企业更重要的工具。文本分析的主要应用之一是在情感分析领域,在该领域,可以从社交媒体信息和博客评论中提取消费者的反馈。文本分析在这一应用中的潜力激发了研发界的研究兴趣。

1)优点

与使用传统数据挖掘工具进行分析相比,将文本分析与传统结构化数据结合可以更全面地了解问题。在情感分析领域应用文本挖掘有助于企业揭露情感,从而改进其客户关系管理(CRM)。文本分析还可以应用于公共安全领域,通过对文本进行审查来确定犯罪活动或恐怖阴谋的特征。

2)问题

文本分析解决方案市场还不成熟。分析引擎将面临非英语内容和地方口语处理方面的挑战。因此,为特定市场开发的文本分析解决方案可能无法直接应用于其他情况(需要一定程度的定制)。文本分析在某些文本源(如技术文档或具有许多特定领域术语的文档)上的适用性也可能存在问题。

采用文本分析不仅是技术部署。了解用于确定结果的度量标准也是一项必备的技能。必须了解要分析什么及如何利用分析结果来改进业务。这需要一定程度的主观性，可能不是管理层所希望的。

2. 内存分析

内存分析是一个分析层，在该层中，详细数据（大小高达太字节）从各种数据源直接加载到系统内存中，以实现快速查询和性能计算。从理论上讲，这种方法部分消除了以关系聚合和预计算数据库的形式构建元数据的需要。

将内存处理作为商业分析的后端资源可以提高性能。在传统的基于磁盘的分析平台上，必须在实际分析过程发生之前创建元数据。元数据的建模方式取决于分析需求。改变元数据的建模方式以满足新的需求需要良好的技术知识水平。内存分析不需要为满足每个终端用户的需求而对元数据进行预建模。分析内容的相关性也得到了改善，因为可以在数据存储在内存中时就对其进行分析。内存分析提供的速度使数据集的交互式可视化成为可能，使数据访问成为一种更具探索性的体验，如图 4-1 所示。

内存分析通过一系列内存技术实现。内存分析技术分类如图 4-2 所示。

内存数据管理：

内存数据库管理系统（IMDBMS）：IMDBMS 将整个数据库存储在计算机 RAM 中，无须使用磁盘 I/O 指令。这使得应用程序可以完全在内存中运行。

内存数据网格（IMDG）：IMDG 提供了一个分布式内存存储，多个分布式应用程序可以在其中存储和检索大量数据对象。

内存低延迟消息传递：

该平台为应用提供了一种可以通过直接内存通信尽快交换消息的机制。

1）优点

内存分析可以处理大型和复杂的数据集，并在几分钟甚至几秒钟内输出大量数据。因此，可以以一种实时或接近实时服务的形式将内存分析提供给用户。由于对元数据的依赖性较低，内存分析启用自助服务分析。当与交互式可视化和数据发现工具结合使用以实现直观、无阻碍和快速的数据浏览时，内存分析可以促进一系列"假设"分析并快速获得结果，有助于通过多次查询迭代对分析结果进行微调，从而做出更好的决策。

内存分析可以用来补充其他大数据解决方案。一种可能是在 Hadoop 平台上运行内存分析，以对分布式数据进行高速分析。NoSQL 数据库可以在内存计算平台上运行，以支持内存分析。内存分析还可以在内存数据网格上运行，该网格可以处理和加工数 PB 的内存数据集。

2）问题

内存分析技术有颠覆企业数据集成工作的潜力。由于该技术支持独立使用，所以企业需要确保他们有办法控制其使用，并且确保从报告到原始源系统的数据沿袭链不间断。

虽然每 GB 内存的平均价格一直在下降，但仍远远高于每 GB 磁盘存储的平均价格。在短期内，与相同数量的磁盘存储相比，内存仍然具有较高的溢价。目前，亚马逊网络服务每 GB 的磁盘存储价格约为 0.10 美元，而内存的价格约为它的 100 倍，即每 GB 10.57 美元。扩展内存分析的使用可能仅限于财力雄厚和技术精明的企业。

3. 预测分析

预测分析是一组统计和分析技术，用于发现大量数据中的关系和模式，这些数据可用于预测行为或事件。预测分析可以挖掘结构化和非结构化数据集及数据流中的信息和模式，以预测未来的结果。预测分析的真正价值在于通过防止影响服务的事件发生，来提供超越传统的被动中断和修复辅助的预测性支持服务，并向主动式支持系统发展。

Gartner 指出，市场上正在评估 3 种用于预测产品支持空间内的技术问题的方法。Gartner 认为，成熟的预测支持服务最终将使用这 3 种方法的组合。

基于模式的方法：该技术试图将实时系统性能和配置数据与非结构化数据源（可能包括已知的故障状况、历史故障记录和客户配置数据）进行比较。在庞大、多层面的存储库中寻找统计模式需要强大的关联引擎，以确定客户的当前配置和性能数据是否指示可能发生的故障。

基于规则的方法：通过对历史性能数据、先前识别的故障模式和应力或载荷测试结果的统计分析，定义了与实时遥测数据进行比较的一系列规则。每项规则可以根据定义的阈值来查询多个遥测数据点和其他外部因素，如一天中的时间、环境条件和并发的外部活动。然后可以检查违反这些规则的情况，并触发升级程序，具体取决于由此产生的问题或停机的严重性和影响。

基于统计过程控制的方法：半个多世纪以来，控制图理论一直是高质量制造过程的支柱，并且已经被证明对管理复杂的过程驱动系统十分有用。具备改造能力的实时遥测技术的出现、数据采集解决方案的改进及支持如此大数据量的网络容量的出现，意味着支撑制造领域质量革命的统计技术现在可以用于 IT 服务的产业化交付。该方法可以轻松地识别统计异常并用于采取适当的应急或预防措施，以确保服务性能不受影响，且业务可以继续正常运行。

1）优点

预测分析利用历史和交易数据中的模式来识别未来的风险和机遇，重点是帮助企业获得可行的见解，并更快地识别和应对新的机会。

预测分析在许多垂直领域都有直接的应用。

执法机构可以寻找犯罪行为和可疑活动的模式，以帮助他们识别可能的动机和嫌疑人，从而更有效地部署人员。例如，埃德蒙顿警察局使用传感器和数据分析来创建定义高犯罪区域的地图，以便主动将更多的警察资源转移到这些区域。

公共卫生当局可以监控不同来源的信息，以寻找某种可能暴发的疾病的症状严重程度。在新加坡，高性能计算研究所（IHPC）开发了一个用于新加坡流行病学研究的仿真模型。在疾病流行的情况下，该模型可用于预测传播情况；税务部门可以使用预测分析来识别传播模式，以突出需要进一步调查的病例。此外，预测分析可用于了解政策对创收的可能影响。

2）问题

由于预测分析需要大量的数据输入，所以面临的挑战在于要知道哪些数据输入是相关的，以及如何集成这些不同的数据源。在一些预测模型中，需要的用户数据触及数据隐私的敏感问题。最终，预测结果的可靠性可能会受到怀疑，特别是当预测结果偏离决策者的观点时。决策者需要时间来接受预测分析的结果并将其作为决策的影响因素。同样，底层预测算法的发展和成熟也需要时间。

4. 基于 SaaS 的商业分析

软件即服务（SaaS）指一个或多个供应商远程拥有、交付和管理软件。一个应用提供一组通用代码，可供许多客户同时使用。基于 SaaS 的商业分析使客户能够快速部署商业分析的一个或多个主要组件，无须大量 IT 参与，也无须部署和维护预置的解决方案。

主要组件如下。

分析应用：通过为特定解决方案提供预打包功能来支持性能管理。

商业分析平台：为开发和集成、信息传递和分析提供环境。

信息管理基础设施：提供数据架构和数据集成基础设施。

1）优点

利用云计算的优势，基于 SaaS 的商业分析提供了一种快速、低成本且易于部署的商业分析解决方案。尤其是对于那些既不具备建立内部分析平台的专业知识，也不打算投资内部商业分析资源的企业来说。对于尚未投资任何形式的内部商业分析解决方案的中小企业来说，基于 SaaS 的商业分析可能非常有用。

2）问题

对于那些希望将数据导出给服务供应商并从服务供应商处提取数据以与内部信息基础架构集成的企业来说，可能存在集成难题。由于数据驻留在云上，基于 SaaS 的商业分析可能不适合需要注意数据隐私和安全问题的企业。

5. 图分析

图分析是对可以转化为由节点和链路组成的图示的数据的研究和分析。图分析非常适用于解决不需要处理数据集中所有可用数据的问题。典型的图分析问题需要用到图遍历技术。图遍历是遍历直接连接的节点的过程。例如，利用图分析可以找出一个社交网络中的两个成员有多少直接和间接连接的方式。

不同形式的图分析如下。

单路径分析：目标是从特定的节点开始，通过图查找路径。先评估可以从起始节点立即到达的所有链路和相应的顶点。根据一组特定的条件从已识别的顶点中选择一个顶点，并进行第一跳。之后继续完成这一过程。结果将得到一条由许多顶点和边组成的路径。

最优路径分析：此分析可以找到两个顶点之间的"最佳"路径。最佳路径可以是最短的路径、价格最低的路径或最快的路径，具体取决于顶点和边的属性。

顶点中心性分析：该分析基于多个中心度评估属性，确定顶点的中心度。

度中心性分析：该分析指示一个顶点有多少条边。边越多，度中心性越高。

亲密中心性分析：该分析标识到其他顶点的跳数最少的顶点。顶点的亲密中心性是指该顶点与其他顶点的接近度。亲密中心性越高，需要到其他顶点的短路径的顶点数越多。

特征向量中心性分析：该分析表明图中顶点的重要性。基于在相同情况下，与高分顶点连接比与低分顶点连接得分更高的原则，将得分分配给顶点。

图分析可用于如下领域。

在金融业，图分析有助于了解资金转移途径。银行账户之间的资金转移可能需要几个中间银行账户，可以应用图分析来确定不同账户持有人之间的不同关系。在庞大的金融交易数据集上运行图分析算法将有助于提醒银行注意可能发生的欺诈性交易或洗钱案件。

在物流业，使用图分析并非新鲜事。最优路径分析是图分析的一种明显形式，可以在物流配送和装运环境中应用。在这一领域有许多使用图分析的例子，其中包括"将货物运送到不同地址的最短路径"和"最划算的货物运送路径"。

社交媒体领域是最现代的图分析应用领域之一。图分析不仅可以用来识别社交网络中的关系，还可以用来理解这些关系。在社交媒体领域使用图分析的结果是从每个社交图表中识别出"有影响力"的人物。然后，企业可以将更多精力用于使这一特定群体参与他们的营销活动或客户关系管理工作。

4.7.2 复杂分析

1. 移动商业分析

移动商业分析是一种新兴趋势，它依赖于移动计算的日益普及。从工作人员的角度来看，需要通过移动商业分析从后端数据存储访问最新的分析见解。移动商业分析有两种类型：被动移动商业分析和主动移动商业分析。

被动移动商业分析围绕"推动"因素展开。基于事件的警告或报告可以在后端刷新后推送到移动设备。被动移动商业分析在为用户提供访问便利性方面可能领先一步，但它不足以支持用户需要的即时分析需求。主动移动商业分析使用户能够与动态的商业分析系统进行实时交互。它可以作为"推"和"拉"技术的结合。对报告的初步看法可以构成"推"，对报告进行进一步分析以获取更多信息的操作可以构成"拉"。

有两种方法可以为移动设备开发商业分析应用：为特定的移动操作系统（如 iOS 或 Android）开发软件，或开发基于浏览器的商业分析应用版本。开发基于浏览器的商业分析应用版本只需要进行一次开发工作，因为可以跨设备开发。另外，定制开发的移动商业分析应用可以提供与设备内容的完全交互，还提供了可以脱机查看数据的定期缓存。

1）优点

移动商业分析有利于进行场外决策。决策者通常只需要访问后端数据存储产生的一些关键分析指标。在移动设备上访问这些指标可以减少决策障碍、提高业务流程效率，并为决策提供更广泛的输入。此外，设备特定功能可以提高移动商业分析的价值。例如，将位置感知与移动商业分析查询结合，可以为查询提供位置上下文，并使用户获得与其位置相关的查询结果。

2）问题

技术和安全风险问题将抑制移动商业分析的应用，特别是在关键任务部署的情况下。如果平板电脑或智能手机设备受到损害，企业的敏感数据可能会面临风险。移动商业分析的发展取决于企业的 BYOD（自带设备）计划的成功。

2. 视频分析

视频分析对数字视频图像进行自动分析，以提取有意义的相关信息，也被称为视频内容分析（VCA）。视频分析使用计算机视觉算法和机器智能来解释、学习和从图像序列中得出推论。视频分析使场景理解自动化，否则将需要人工监控。它不仅可以检测运作，而且可以将运作限定为对象，了解对象的上下文并通过场景跟踪对象。从理论上讲，任何可以在视频图像上看到并准确定义的行为都可以被自动识别，既而触发恰当的响应。

视频分析架构包含两种系统，即基于边缘的系统和基于中心的系统。

1）基于边缘的系统

基于边缘的系统如图 4-3 所示。在该系统中，会在应用任何压缩之前对视频源（即摄像机）上的原始图像进行视频分析。这意味着图像具有最大数量的信息内容，可以使分析更有效。对本地数据的处理使系统能够应对传输或网络故障。恢复网络连接后，警报可以在本地响应，也可以在恢复网络连接后存储以进行传输。在没有正在处理的事件或警报时，可以通过降低帧率、降低分辨率和提高图像压缩率来控制带宽占用。基于边缘的系统必须通过具有足够处理能力的 IP 摄像机或视频编码器来实现。与传统的模拟摄像机或网络摄像机相比，这可能会导致更大的设备具有更高的功率要求。

2）基于中心的系统

基于中心的系统如图 4-4 所示。在该系统中，视频分析是通过专用服务器实现的，该服务器可以提取视频、分析视频并发布警报或分析结果。由于分析设备安装在一个中心位置，所以便于维护和升级。使用中央不间断电源（UPS）可以防止系统断电。虽然这种独立于摄像机的方法适用于大多数类型的监视系统，但是可能需要大量的网络带宽才能从图像采集设备传输高质量的图像。由于图像中的信息丢失，压缩和传输效果可能会影响分析的效率和准确性。

3）优点

视频分析可以用安全意识来补充商业智能。具有分析功能的自动监控摄像头可以检测到人员和车辆的存在，并解读其活动。游荡或进入未经授权的区域等可疑活动会被自动标记并转发给安全人员。在零售业，视频分析可以通过更好地了解客户体验来提升客户体验并推动创收。通过视频分析，零售商可以分析顾客在商店的时间，以评估关键店促销区的有效性。零售商还能够通过将销售收入与客户的人口统计资料相关联来确定商店访客的人口统计特征，并识别购买模式。

4）问题

尽管技术的进步改进了场景和对象识别等的视频功能，但是当前的分析仍然集中在入侵检测和人群管理上。在这种情况下，事件随机发生且通常会吸引到需要定制的特定垂直领域。此外，视频分析仍然可能被准确性和可靠性问题困扰，包括阴影和光线下高对比度区域造成的问题。视频分析系统可能会出现误报，主要由分析算法的复杂程度导致。误报率（FAR）是指当什么都没发生时，系统发出警报的次数。每台摄像机每天发出一个误报听起来是可以接受的，但在成百上千台摄像机叠加后，这个数字可能会非常大。例如，在一个由100台摄像机组成的网络中，上述情况相当于每14分钟发出一次误报。

硬件限制是一个问题。虽然视频分析可以在设备的边缘进行，但是每台摄像机的部署成本或硬件需求，以及安装在每台摄像机上的用于运行复杂视频分析算法的计算容量将非常大。另外，在中央服务器上进行的视频分析需要高带宽，而用于视频分析的图像馈送可能会受衰减和细节丢失的影响，进而影响视频分析的准确性。基于IP的摄像机正在兴起，但目前的大多数基础设施仍在使用基于模拟的摄像机。这两种技术的划分也会导致采用视频分析的成本增加。

3. 音频分析

音频分析使用一种通常被称为音频挖掘的技术，这种技术会搜索大量音频数据以获取特定的音频特征。当应用于语音识别领域时，音频分析识别音频中的单词并将它们放入搜索文件中。最常用的两种音频挖掘方法是大词汇量连续语音识别和语调识别。

大词汇量连续语音识别：大词汇量连续语音识别（LVCSR）将语音转换为文本，然后使用词典来理解所说的内容。词典通常包含几十万个词条，其中包括通用词及行业和企业的特定术语。分析引擎可以使用词典来处理音频的语音内容，以生成可搜索的索引文件。索引文件包含与它在音频数据中理解的单词有关的信息，可以快速搜索关键词和短语，以显示包含搜索词的相关对话。

语调识别：语调识别不需要从语音到文本的任何转换，只对声音起作用。分析引擎首先分析和识别音频内容中的声音，以创建基于语调的索引。然后使用包含几十个音素的词典将搜索词转换为正确的音素字符串。最后，系统在索引中查找搜索项。

语调识别和LVCSR的区别之一是哪个阶段（索引或搜索）的计算最集中。使用语调识别的音频内容索引速度比使用 LVCSR 快很多倍。然而，在搜索阶段，与 LVCSR 相比，语调识别系统的计算负担较大，而搜索过程通常较为简单。语调识别不需要使用复杂的语言模型，因为它可以有效运行，而不需要知道哪些单词是预先识别的。相反，在 LVCSR 系统中，知道哪些单词是预先识别的对于获得良好的识别精度来说至关重要。因此，LVCSR 方法必须使用复杂的语言模型，从而在索引阶段产生更大的计算量，并使索引速度显著降低。

语调识别方法的优点是可以保持开放的词汇表，这意味着不需要重新处理音频即可搜索个人或企业名称。在 LVSCR 系统中，任何在索引语音时系统不知道的单词都永远无法找到。例如，一个以前不知道的术语"消费主义"现在很流行。LVSCR 系统使用的词典中没有这个单词，这意味着分析引擎永远无法在系统处理的音频文件中找到这个单词。为了找到这个新单词，LVSCR 系统必须用包含"消费主义"一词的新词典进行更新，并且必须对所有音频再次进行预处理，这个过程十分耗时。语调分析系统则不会出现这个问题，因为它工作在音素级别，而不是单词级别。只要在搜索时可以生成单词的语调，就可以找到匹配的单词，无须重新处理音频。

1）优点

以语音分析形式使用的音频分析使拥有电话服务中心的企业能够获得市场情报。电话服务中心维护的通话记录通常代表数千小时的"客户之声"，代表企业将要提取和挖掘的市场洞察力。语音分析可以用于分析企业的所有通话记录，以便从尽可能广泛的来源收集数据。更重要的是，数据是从实际的客户交互中收集的，而不是从以后的交互中收集的。因此，在实际上下文中的数据更加准确。

为了安全和获取情报，政府和监管机构必须处理大量对话记录。手工抄写和分析不仅速度慢，还会面临迟的或错过情报的风险。音频分析可以用于分析实时音频流，用户可以将直接移动到对话记录中使用感兴趣的单词或短语的特定位置，从而改进监控过程。

2）问题

音频分析的准确性和可靠性会受一些限制。音频流的识别很难达到 100% 的准确率。在语音分析的情况下，系统可能无法处理重音词。此外，即使分析算法足够复杂，能够高精度地识别单词，理解单词的上下文含义也是一项挑战。

Unit Five Big Data Application

5.1 Learning Goals

After learning this unit, you will be able to master the following knowledge:

1. Big Data has a huge application in healthcare, particularly in areas where analysis of large data sets is a necessary precondition for creating value.

2. Top retailers are mining customer data and using Big Data technologies to help make decisions about their marketing campaigns, merchandising and supply chain management.

3. As more emphasis is being placed on learning that is adaptive, personal and flexible, there is the need to mine unstructured data such as student interactions and any form of student-generated content.

4. Big Data analytics offers the opportunity for public transport service operators to obtain critical insights on passenger demand trends so as to implement more effective measures in their service provisions.

5. Big Data plays a significant role in the finance sector, especially with regard to fraud detection with the application of Complex Event Processing.

6. The harvesting of large data sets and the use of analytics clearly involve data privacy and security concerns.

5.2 Dialogue

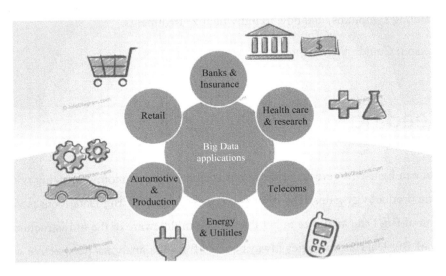

Because Big Data is so unique in the history of the world. Its content and applications keep improving. Many applications about it have surfaced. Henry, Mark, Sophie and Professor Jackson are having a discussion about it.

Prof Jackson: With the enhanced capabilities of data storage and rapid computation as well as real-time delivery of information via the Internet, the average daily consumption of data by an individual has grown exponentially. Let's take our school for example. Students are using an online system to meet the requirements of the individual learning materials. Meanwhile, teachers are using this system to collect and record the students' online learning behaviour, which is saved into the database. Via data analysis and processing, teachers can predict the students' academic performance and give students guidance and help in time. Can you illustrate more examples in other fields?

Henry: Definitely! Big Data can help power companies and their customers conserve energy by giving feedback messages from a malfunctioning meter to the power company or alarming customers when they are wasting energy.

Prof Jackson: Marvelous!

Mark: Big Data analytics is helping police departments to develop prediction algorithms to prevent probable crime commitments.

Sophia: Publishing houses use data from text analytics and Social Networks to give readers personalized news and transform a large amount of news simultaneously into native languages.

Henry: As far as I'm concerned, health care is one of the biggest opportunities. If we had electronic records of Americans going back generations, we'd know more about genetic propensities, correlations among symptoms, and how to individualize treatments.

Prof Jackson: Couldn't agree more!

5.3　Reading

Big Data can be used to create value across sectors of the economy, bringing with a wave of innovation and productivity gains. The discussion on the impact of Big Data focuses very much on the application of Big Data analytics rather than on the middleware or the infrastructure. Therefore, the adoption of Big Data technologies always comes from the analytics perspective which in turn drives the adoption of the underlying supporting technologies. According to McKinsey, there are five ways to leverage Big Data's potential to create value.

Creating transparency: making Big Data more accessible to relevant stakeholders across disparate departments in a timely manner can create value by sharply reducing the data search and processing time.

Enabling experimentation: as organizations create and store more transactional data in digital form, they can collect more accurate and detailed performance data on many relevant aspects, such as product inventories and staff movements. This data can be used to analyse variability in performance, identify root causes and discover needs or opportunities.

Segmenting populations: organizations can leverage Big Data technologies to create highly specific customer segmentations and to customize products and services that meet there needs. Though this functionality may be well-known in the field of marketing and risk management, its use in other sectors, particularly in the public sector, is not common and to a certain extent, may be considered revolutionary.

Replacing or supporting human decision making: sophisticated analytics with automated algorithms can unearth valuable insights that would otherwise remain hidden. These insights can then be used to minimize decision risks and improve decision making. In some cases, decisions may not be completely automated but only augmented by the analysis of huge data sets using Big Data techniques rather than small data sets and samples that individual decision makers can handle and understand.

Innovating new business models, products and services: using emerging Big Data technologies, companies can enhance and create new products and services.

The application of Big Data varies across verticals because of the different challenges that bring about the different use cases. The common driver of Big Data analytics across verticals is to create meaningful insights that translate to new economic value. Adoption of Big Data analytics can be seen in the following verticals.

5.3.1 Healthcare

Big Data has a huge potential in the application of healthcare, particularly in areas where analysis of large data sets is a necessary precondition for creating value. Possible adoption of Big Data analytics could be done in a few specific areas. One of them is Comparative Effectiveness Research (CER). CER is designed to inform healthcare decisions by providing evidence on the effectiveness, benefit and harm of different treatment options. The evidence is generated from research studies that compare drugs, medical devices, tests, surgeries, or ways to deliver healthcare. By analysing large data sets that include patient genome characteristics, and the cost and outcomes of all related treatments, healthcare services can identify the most clinically effective and cost-effective treatments. However, before any analytical techniques can be used, comprehensive and consistent clinical data sets must be captured, integrated and made available to researchers. In this area, issues of data standards and interoperability, as well as patient privacy, conflict with the provision of sufficiently detailed data to allow effective analysis.

The other application of Big Data analytics can be in the area of a Clinical Decision Support System (CDSS). CDSS is a computer application that assists clinicians in improved decision making by providing evidence-based knowledge with respect to patient data. The system analyse physician entries on patient data and compare them against medical guidelines to alert for potential errors such as adverse drug reactions, in the process reducing adverse reactions and resulting in lower treatment error rate that arise from clinical mistakes. The system can be extended to include the mining of medical literature to create a medical expert database capable of suggesting treatment options to physicians based on patient records. The system can also be extended to include image analytics to analyse medical images such as X-rays and CT scans for pre-diagnosis.

Healthcare service quality can be improved using Big Data analytics. The US Centres for Medicare and Medicaid Services (CMS) announced new data and information initiative to be administered by the Office of Information Products and Data Analytics (OIPDA). The initiative will guide the agency's evolution from a "fee-for-service" based payer to a "value-based purchaser of care" and aims to link payments to the quality and efficiency of care rather than the sheer volume of services. To achieve this objective, there is a need to analyse data collected from service providers such as hospitals and physicians across the country to measure the service quality so as to promote the establishment of standards in the healthcare services sector. Big Data solutions may effectively meet the challenges faced by CMS and similar organizations around the world.

Predictive analytics can be applied to analyse patient profiles to identify individuals who are liable to contract a particular disease. Proactive treatments can be administered in an effort to prevent illness or limit its severity so as to control healthcare costs. The ability to identify patients most in need of services has implications for improved treatment quality and financial savings.

5.3.2 Retail

The retail sector is built on an understanding of consumers' retail habits. Top retailers are mining customer data and using Big Data technologies to help make decisions about their marketing, merchandising and supply chain management. Retailers are using more advanced methods in analysing the data they collect from multiple sales channels and interactions. The use of increasingly granular customer data gives retailers access to more detailed analytics insights which, in turn, can improve the effectiveness of their marketing and merchandising efforts. The clearer insights will also provide retailers greater accuracy in forecasting stock movements, thereby improving supply chain management.

1. Marketing

There are many ways which Big Data analytics can be applied in retail marketing. One of these applications is to enable cross-selling which uses all the data that can be known about a customer, including the customer's demographics, purchase history and preferences, to increase the average purchase amount. Online retailer, Amazon, is a good example. With its "You may also like" recommendation engine, Amazon is able to provide product recommendations relevant to each of its customers, based on their browsing, shopping habits and other online profiles. According to the company, 30% of its sales are generated by its recommendation engine.

Another application would be for the purpose of analysing customers' in-store behaviour so as to help retailers create a more conducive shopping environment and increase the customers' propensity to purchase. A possibility here is to adopt video analytics to analyse customers' in-store traffic patterns and behaviour to help improve store layout, product mix and shelf positioning. In addition, Big Data analytics can be used to harness consumer sentiments. Sentiment analysis leverages the streams of data generated by consumers on various social media platforms to help fine-tune a variety of marketing decisions. Big Data analytics helps retailers create micro-segments of their customer pool. Along with the increase in customer data, the increasing sophistication in analytics tools has enabled more granular customer profiling, helping retailers to provide more relevant, if not personalized, products and services.

2. Merchandising

Retailers have to consider the optimization of their product assortments. Deciding the right product mix to carry in the stores based on analysis of local demographics, buyer perception and purchasing habits can increase sales. Retailers can use the increasing granular pricing and sales data together with the analytics engine to optimize their pricing strategy. Complex demand and elasticity analytics models can be used to examine historical sales data to derive insights into product pricing at the Stock Keeping Unit (SKU) level. Retailers can also use the data and consider future promotion events to identify causes that may drive sales.

3. Supply Chain Management

Inventory management is crucial in any successful retail strategy. With the details offered by Big Data analytics from the multiple huge data sets, retailers can have a clear picture of their stock movements and improve on their inventory management. A well-executed inventory management strategy allows retailers to maintain low stock levels because the suppliers respond more accurately to consumer demands. Retailers would want to optimize their stock levels so as to reduce inventory storage costs while minimizing the lost sales due to merchandise stock-outs.

5.3.3 Education

In the education sector, learners are creating information at the same time as they are consuming knowledge. Students are faced with increasingly demanding curricula where they are no longer expected to regurgitate facts from hard memorizing but are required to learn the subjects with deep understanding. At the same time, the onus is also on the educators as high expectations are placed on them to provide personalized teaching and mentoring. The challenge for educators is to be able to have a clear profile of each student under their charge. Creating a profile for each student would require disparate sets of information and this is where the opportunity lies for Big Data analytics.

As more emphasis is being placed on learning that is adaptive, personal and flexible, there is the need to mine unstructured data such as student interactions and any form of student generated content. Learning analytics tool, such as Social Networks Adapting Pedagogical Practice (SNAPP), can be deployed to analyse this data. SNAPP is a software tool that allows users to visualize the network of interactions resulting from discussion forum posts and replies. The visualization provides an opportunity for teachers to rapidly identify patterns of user behaviour at any stage of course progression. This information provides quick identification of the levels of engagement and network density emerging from any online learning activities. From there, disengaged and low performing students can be identified. An example of a network visualization diagram is shown in Figure 5-1.

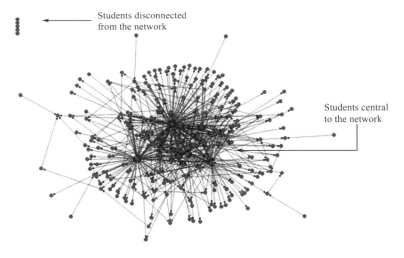

Figure 5-1　An example of a network visualization diagram.

Other applications for Big Data analytics can be exemplified by Civitas Learning, a digital education platform that uses predictive analytics to help guide educational decision making. Civitas Learning takes data such as demographics, behavioural and academic data provided by its partner institutions, and anonymizes and combines them. The data is then analysed to identify trends and provide insights, such as identifying the courses and tracks that are most beneficial, based on the student profiles and the instructional approaches that tend to be most effective at ensuring good educational outcomes. These insights are then translated into real-time recommendations for students, instructors, and administrators through a customized platform.

5.3.4 Transport

1. Personalized Real-time Information for Travel Options

The prevalence of smartphones makes it possible for individuals to have information tools that identify and provide a variety of travel options, enabling commuters to make the best decision in terms of time and costs. By entering a destination, a commuter could be given the estimated time of arrival for different travel options that include buses, trains or private car transport. This requires feeding data from sensors placed on board the public transport vehicles as well as along the roads into computation models that predict traffic patterns and travel times at different times of the day. The model can also be built to take into account other factors, such as the likelihood of a crash on a particular roadway at a given time of day, weather conditions and even the anticipated fuel consumption and cost. These models can enable predictions of train and bus arrival times, and compare these mass transit approaches with different routes the commuter could take by car, and make recommendations of how to travel to a destination in the most effective way.

2. Real-time Driver Assistance

Analytics of the data from Sensor Networks can provide information that could help drivers navigate the road. This includes the incorporation of real-time road condition information to enable active routing. Drivers will always be advised in real-time to use the least busy roads. As the sophistication of the algorithm increases, the re-routing of vehicles will include the avoidance of road closures and roadworks. This approach can optimize road usage in a more balanced manner, thereby mitigating traffic congestion and decreasing the overall travel time without necessitating expensive new transport networks.

3. Scheduling of Mass Transit System

Analysis of passenger boarding data, bus and train location data, and a series of Sensor Networks data could allow for accurate counts of the number of people currently using the system and the number of people waiting at each stop. This information could be used to dynamically manage the vehicles, based on actual demand.

4. Preventive Maintenance

Preventive maintenance is a major cost saver for organizations with significant infrastructure assets and regular maintenance schedules. A regular maintenance schedule is a basic requirement for any transport service provider to ensure a serviceable transport fleet (e.g., taxi or bus). However, despite the best maintenance effort, there will still be cases of vehicle break-downs which cause disruptions to the transport services. The maintenance effort is not universal across all vehicles in the transport fleets, certain vehicles would need more maintenance efforts compared to others. It is difficult to identify the specific vehicles that would need more maintenance efforts. However, by applying data analytics, transport service operators will be able to predict issues ahead of outages and improve client satisfaction. By capturing the log data from the vehicles onto an analytics platform, algorithms can be executed to report issues and forecast events related to each vehicle in the fleet. Preventive maintenance, rather than regular maintenance, is more effective in ensuring the transport fleet's serviceability.

5. Improved Urban Design

The enhancement of transport system includes improvements to urban design. Urban planners can more accurately plan for road and mass transit construction and the mitigation of traffic congestion by collecting and analysing data on local traffic patterns and population densities. With the help of sensors and personal location data, urban developers can have access to information about peak and off-peak traffic hotspots, and volumes and patterns of transit use. The information can help urban planners make more accurate decisions on the placing and sequencing of traffic lights, as well as the likely need for parking spaces.

5.3.5　Finance

Big Data plays a significant role in the finance sector, especially with regard to fraud detection with the application of Complex Event Processing (CEP). By relating seemingly unrelated events, CEP aims to give companies the ability to identify and anticipate opportunities and threats. CEP is typically done by aggregating data from distributed systems in real-time and applying rules to discern patterns and trends that would otherwise go unnoticed. It is with the analysis of these huge data sets that fraud activities can be more easily detected. For example, unusual spending patterns such as buying French train tickets online from a US IP address minutes after paying for a restaurant bill in China. Nonetheless, this is only possible when IT systems of large financial institutions are able to automatically collect and process a large volume of data from an array of sources including Currency Transaction Reports (CTRs), Suspicious Activity Reports (SARs), Negotiable Instrument Logs (NILs), and Internet-based activity and transactions. It would be ideal to include the entire history of profile changes and transaction records to determine the rate of risk for each of the accounts, customers, counterparties, and legal entities, at various levels of aggregation and hierarchy. While this was traditionally impossible due to constraints in processing power and cost of storage, HDFS has made it possible to incorporate all the detailed data points to calculate such risk profiles and have the results sent to the CEP engine to establish the basis for the risk model. A Database Management System will capture and store low-latency and a large volume of data from various sources, as well as real-time data integration with the CEP engine, to enable automatic alerts and trigger business processes to take appropriate actions against potential fraud.

Companies in the finance sector also apply Big Data analytics to understand customer behaviour in order to increase customer intimacy and predict the right product to introduce at the appropriate time. This involves understanding customers and competitors, and using computational algorithms to make sense of the world. High performance analytics can help to reach customers at precisely the right time and place, and with the right message, so banks can acquire and grow a profitable customer base. Any companies such as British Pearl, a technology-based financial solutions company, make use of unstructured data to find the best ways of attracting and retaining customers. Since it costs much more to acquire a customer than maintain one, a large financial services firm in USA uses data from 17 million customers and 19 million daily transactions as an early warning system to detect customer disengagement. Certain interactions and transactions trigger alerts to front-line staff who immediately contact the customer whenever there is an indication that the relationship needs to be nurtured. Not only do finance companies use their own data sets, they also work with partners operating in and out of the financial sector to get a far more comprehensive and accurate view of the market. By obtaining data about potential customers and their online presence, products and services can be tailored more accurately to specific individuals, and customers can be more effectively retained.

Working with various data sources can help financial institutions identify patterns and trends which measure people's likelihood to be fraudulent in order to reduce the bank's home loan lending risk. Getting insights into consumers' major life events reduces the bank's risk exposure across the customers' lifecycles. To assess the creditworthiness and conduct risk evaluation, banks can apply high performance analytical techniques to detect subtle anomaly throughout large loan portfolios.

Banks can use these techniques to reduce the time needed to identify problem loans from more than 100 hours to a few minutes, resulting in significant savings.

5.3.6 Big Data Privacy and Security

The systematic approaches toward data collection in order to enhance randomness in data sampling and reduce bias are not apparent in the collection of Big Data sets. Big Data sets do not naturally eliminate data bias. The data collected can still be incomplete and distorted which, in turn, can lead to skewed conclusions. Consider the case of Twitter which is commonly scrutinized for insights about user sentiments. There is an inherent problem with using Twitter as a data source as only 40% of Twitter's active users are merely listening and not contributing. This may suggest that the tweets come from a certain type of people (probably people who are more vocal and participative in social media) than from a true random sample. In addition, Twitter makes a sample of its materials available to the public through its streaming Application Programming Interface (API). It is not clear how the sample of materials is derived.

The harvesting of large data sets and the use of analytics clearly involve data privacy and security concerns. The tasks of ensuring data security and protecting privacy become harder as the information can easily transcend geographical boundaries. Personal data such as health and financial data can help to bring about significant benefits such as helping identify the right medical treatment or the appropriate financial services or products. Likewise individual shopping and location information can help to bring about better shopping experience by informing customers of the products and services that are of greater relevance.

Organizations may have tried to use various methods of de-identification to distance data from the real identities and allow analysis to proceed while at the same time containing privacy concerns. However, researchers have proven that anonymized data can be re-identified and attributed to specific individuals. These research outcomes disrupt the privacy policy landscape by undermining the faith that can be placed in anonymization. The implications for government and businesses can be stark, given that de-identification has become a key component of numerous business models, more notably in the contexts of targeted marketing and health data.

Note:

The text is adapted from: https://www.studymode.com/essays/Bigdata-39129403.html.

5.4　New Words & Phrases & Sentences

5.4.1　New Words

1. merchandising　　　　　　　　　*n.* 推销；展销；衍生产品；相关产品的销售
2. transparency　　　　　　　　　*n.* 幻灯片；透明正片；透明；透明性，透明度；显而易见；一目了然
3. leverage　　　　　　　　　*n.* 影响力；杠杆作用；杠杆效力
　　　　　　　　　v. 利用；举债经营；借贷收购
4. clinically　　　　　　　　　*adv.* 临床
5. integrate　　　　　　　　　*v.* （使）合并，成为一体；（使）加入，融入群体
6. Medicare　　　　　　　　　*n.* 医疗保障制度
7. fraudulent　　　　　　　　　*adj.* 欺诈的，欺骗性的
8. transcend　　　　　　　　　*vt.* 超越，胜过
9. rapidly　　　　　　　　　*adv.* 迅速；高速；急速地；急促
10. disengage　　　　　　　　　*v.* （使）脱离，松开；解脱；（使）停止交战，脱离接触
11. systematic　　　　　　　　　*adj.* 成体系的；系统的；有条理的；有计划、有步骤的
12. demographic　　　　　　　　　*n.* 人口统计数据，人口统计资料；（尤指特定年龄段的）人群
　　　　　　　　　adj. 人口学的；人口统计学的
13. academic　　　　　　　　　*adj.* 学业的，教学的，学术的
14. prevalence　　　　　　　　　*n.* 流行；卓越

15. sophistication   *n.* 世故；复杂巧妙；高水平
16. browse   *v.* （在商店里）随便看看；浏览；翻阅；（在互联网上）搜寻信息，浏览信息
17. conducive   *adj.* 使容易（或有可能）发生的；有益的
18. propensity   *n.* （行为方面的）倾向；习性
19. harness   *n.* 马具；挽具；背带，保护带
        v. 给（马等）上挽具；控制，利用；用挽具把……套到……上
20. optimize   *v.* 使最优化；充分利用
21. anonymize   *v.* 消除（实验结果的）对象信息；使匿名化
22. commuter   *n.* （远距离）上下班往返的人，通勤者
23. dynamically   *adv.* 动态地；充满活力地；不断变化地

5.4.2 Phrases

1. Complex Event Processing (CEP)   复杂事件处理
2. the Office of Information Products and Data Analytics (OIPDA)   信息产品和数据分析办公室
3. Segmenting populations   细分人群
4. Social Networks Adapting Pedagogical Practice (SNAPP)   社交网络自适应教学实践工具
5. preventive maintenance   预防性维护
6. replacing or supporting human decision making   替代或支持人工决策
7. cost-effective treatment   划算的治疗方法
8. sophisticated analytics   复杂分析
9. Comparative Effectiveness Research (CER)   比较效益研究

10.	Clinical Decision Support System (CDSS)	临床决策支持系统
11.	be applied in	应用于
12.	multi-dimensional space	多维空间
13.	proactive treatments	预防性治疗
14.	the average purchase amount	平均购买量
15.	micro-segments of their customer pool	客户群的细分市场

5.4.3 Sentences

1. Big Data can be used to create value across sectors of the economy, bringing with a wave of innovation and productivity gains. 大数据可以用来在各经济部门创造价值，并带来一波创新和生产率提高。

2. The discussion on the impact of Big Data focuses very much on the application of Big Data analytics rather than on the middleware or the infrastructure. 关于大数据影响的讨论主要集中在大数据分析的应用方面，而不是中间件或基础设施方面。

3. In some cases, decisions may not be completely automated but only augmented by the analysis of huge data sets using Big Data techniques rather than small data sets and samples that individual decision makers can handle and understand. 在某些情况下，决策可能不是完全自动化的，只能通过使用大数据技术而不是单个决策者能够处理和理解的小数据集和样本对大数据集进行分析来增强。

4. Using emerging Big Data technologies, companies can enhance and create new products and services. 利用新兴的大数据技术，企业可以增强和创造新的产品和服务。

5. The common driver of Big Data analytics across verticals is to create meaningful insights that translate to new economic value. 大数据分析在垂直领域的共同动力是创造转化为新的经济价值的有意义的见解。

6. Big Data has a huge potential in the application of healthcare, particularly in areas where analysis of large data sets is a necessary precondition for creating value. 大数据在医疗保健行业有巨大的应用潜力，特别是在那些将分析大型数据集作为创造价值的必要先决条件的领域。

7. The system analyse physician entries on patient data and compare them against medical guidelines to alert for potential errors such as adverse drug reactions, in the process reducing adverse

reactions and resulting in lower treatment error rate that arise from clinical mistakes. 该系统分析医生输入的患者数据，并将其与医疗指南进行比较，以对潜在的问题发出警告（如药物不良反应），从而减少不良反应并降低由临床错误导致的治疗出错率。

8. The system can be extended to include the mining of medical literature to create a medical expert database capable of suggesting treatment options to physicians based on patient records. 可以将该系统扩展为具有挖掘医学文献功能的系统，从而创建一个医学专家数据库，该数据库能够根据患者记录向医生提出治疗方案建议。

9. To achieve this objective, there is a need to analyse data collected from service providers such as hospitals and physicians across the country to measure the service quality so as to promote the establishment of standards in the healthcare services sector. 为了实现这一目标，需要通过分析从全国各地的医院和医生等服务提供者处收集的数据来衡量服务质量，以促进医疗服务行业标准的建立。

10. There is an inherent problem with using Twitter as a data source as only 40% of Twitter's active users are merely listening and not contributing. 使用 Twitter 作为数据源存在一个固有的问题，因为有 40% 的 Twitter 活跃用户只是在接收信息，而并不发出信息。

5.5 Exercises

【Ex. 1】Content Questions.

1. Which five ways to leverage Big Data's potential to create value?

2. What is the role of Big Data in healthcare?

3. What is the role of Big Data in marketing?

4. What are the roles of Big Data in transport?

【Ex. 2】句子翻译。

1. In this case, you can create a data source that retrieves data from the work items in the repository.

2. This is a quick way to have a form that you can edit and tailor according to your requirements.

3. An attacker who successfully exploited this vulnerability could run arbitrary code as the

logged-on user.

4. We are using this transistor to amplify a telephone signal.

5. The data set might also contain another table with order information.

6. This establishes a workflow between use cases.

7. Consequently, each registered base node might have different user registries configured if security is enabled.

8. Older machines will need a software patch to be loaded to correct the date.

9. Administrative staff may be deskilled through increased automation and efficiency.

10. At this point, you may activate or deactivate whatever other plugins you wish.

【Ex. 3】将下列词填入适当的位置（每词只用一次）。

| monitoring | exposure | relieves | reputational | mobile |
| requirements | organization | storage | independently | growing |

Key Challenges for Big Data Security

Cyber Criminals. As it becomes bigger and more difficult to manage, Big Data consequently becomes more appealing to hackers and cyber criminals. Because Big Data is a data set of unprecedented size with centralized access, any __(1)__ is total exposure. These types of breaches make headlines, incite consumers, and may cause major __(2)__, legal, and financial damage.

Resource Capacity. As an organization collects Big Data across channels at an exponential rate, their __(3)__ can grow beyond TeraBytes. As a result, data encryption and migration can get bottlenecked or leaky. Additionally, the sheer volume of data makes the implementation of security control unwieldy. The tools required for __(4)__ and analysing Big Data produce massive amounts of their own security-related data every day, which puts undue pressure on the organization's capacity to store and analyse it all.

Cloud and Remote Access. One answer to the capacity issues of Big Data is to put it in the cloud. This __(5)__ some of the burden for storage and processing, but creates new challenges for protecting it from criminals. And as more businesses allow for flex-time and __(6)__ offices, employees have access to sensitive company data via smartphones, tablet devices, and home laptops. Protecting personal devices becomes a balancing act between security and productivity.

Supply Chain and Partner Security. Organizations rarely operate (7). They rely on supply chain partners and external vendors for many of their business functions. Information flows in and out of each (8) to keep these relationships functioning. Coordinating the safety of Big Data across partners is another layer of complexity to a business's information security challenges.

Privacy. Both private and public organizations face the (9) challenge of privacy concerns. Consumers are wary about personal information being collected and stored, and fearful about security breaches. Plus, there are legislative and regulatory (10) to keep in mind.

【Ex. 4】补充空白部分的单词。

1. Big Data can be used to _____ _____ across sectors of the economy, bringing with a wave of _____ and _____ _____.

2. The _____ on the impact of Big Data _____ very much on the application of Big Data analytics rather than on the _____ or the _____.

3. This data can be used to _____ _____ in performance, identify root causes and discover _____ or _____.

4. Though this functionality may be well-known in the field of _____ and _____ _____, its use in other sectors, _____ in the public sector, is not common and to a certain extent, may be considered _____.

5. _____ _____ _____, decisions may not be completely automated but only augmented by the analysis of huge data sets using _____ _____ _____ rather than small data sets and samples that individual decision makers can handle and understand.

6. The common driver of Big Data analytics across verticals is to _____ _____ _____ that translate to new economic value.

7. Big Data has a huge potential in the application of _____, particularly in areas where analysis of large data sets is a necessary precondition for _____ _____.

8. The evidence is generated from research studies that compare _____, _____, _____, _____, _____, or ways to deliver healthcare.

5.6　Translations for Dialogue

　　因为大数据在世界历史上是独一无二的，所以它的内容和应用持续改善，已经出现了许多应用。亨利、马克、索菲和杰克逊教授正在讨论这个问题。

　　杰克逊教授：随着数据存储能力和快速计算能力的提高及互联网实时信息传递功能的增强，每人每天的平均数据消费量呈指数级增长。以我们学校为例，学生使用在线系统来满足对个人学习资料的需求。同时，教师利用该系统收集和记录学生的在线学习行为，并将其保存到数据库中。通过数据分析和处理，教师可以预测学生的学习成绩，及时给予学生指导和帮助。你能举其他领域的例子进行说明吗？

　　亨利：当然可以！大数据可以通过向电力公司提供故障电表的反馈信息或在客户浪费能源时提醒客户，来帮助电力公司及其客户节约能源。

　　杰克逊教授：好极了！

　　马克：大数据分析正在帮助警察局开发预测算法，以防止可能发生的犯罪行为。

　　索菲：出版社利用文本分析和社交网络的数据为读者提供个性化新闻，同时将大量的新闻转化为读者的母语。

　　亨利：我认为，医疗保健是最大的机遇之一，如果我们有上一代美国人的电子记录，我们就能更多地了解遗传倾向、症状之间的相关性，以及如何实现个体化治疗。

　　杰克逊教授：我完全同意你的看法！

5.7　Translations for Reading

　　大数据可以用来在各经济部门创造价值，并带来一波创新和生产率提高。关于大数据影响的讨论主要集中在大数据分析的应用方面，而不是中间件或基础设施方面。因此，大数据技术的应用总是从分析的角度出发，反过来又推动了底层支撑技术的应用。麦肯锡认为，有5种方法可以利用大数据的潜力创造价值。

　　创建透明度：使不同部门的利益相关者能够及时访问大数据，可以通过大大缩短数据搜索和处理时间来创造价值。

进行实验：随着企业以数字形式创建和存储更多的交易数据，它们可以在许多相关方面（如产品库存和员工流动）收集更准确和详细的绩效数据。这些数据可用于分析性能的可变性、确定根本原因及发现需求或机会。

细分人群：企业可以利用大数据技术创建非常具体的客户细分，并定制满足其需求的产品和服务。虽然这一功能在市场营销和风险管理领域可能是众所周知的，但它在其他行业（尤其是在公共行业）的使用并不常见，并且在某种程度上可以认为它是革命性的。

替代或支持人工决策：使用自动算法的复杂分析可以挖掘出有价值的见解，否则这些见解将继续隐藏起来。这些见解可用于最小化决策风险并改善决策。在某些情况下，决策可能不是完全自动化的，只能通过使用大数据技术而不是单个决策者能够处理和理解的小数据集和样本对大数据集进行分析来增强。

创新商业模型、产品和服务：利用新兴的大数据技术，企业可以增强和创造新的产品和服务。

带来不同用例的不同挑战使大数据在不同垂直领域的应用不同。大数据分析在垂直领域的共同动力是创造转化为新的经济价值的有意义的见解。大数据分析应用的垂直领域如下。

5.7.1 医疗保健

大数据在医疗保健行业有巨大的应用潜力，特别是在那些将分析大型数据集作为创造价值的必要先决条件的领域。在几个特定领域可以采用大数据分析。其中之一是比较效益研究（CER）。CER 旨在通过提供关于不同治疗方案的有效性、益处和危害的证据，为医疗保健决策提供依据。这些证据来自对药物、医疗器械、测试、手术或提供医疗保健的方式进行比较的研究。通过分析包含患者基因组特征、所有相关治疗成本和治疗效果的大型数据集，医疗健康服务可以确定临床上最有效和最划算的治疗方法。然而，在使用任何分析技术之前，必须先收集、整合并向研究人员提供全面一致的临床数据集。在这一领域，数据标准、互操作性及患者隐私问题与提供足够详细的数据以进行有效分析相冲突。

大数据分析的另一个应用是在临床决策支持系统（CDSS）领域。CDSS 是一种计算机应用程序，可以通过提供与患者数据有关的循证知识，来帮助临床医生改进决策。该系统分析医生输入的患者数据，并将其与医疗指南进行比较，以对潜在的问题发出警告（如药物不良反应），从而减少不良反应并降低由临床错误导致的治疗出错率。可以将该系统扩展为具有挖掘医学文献功能的系统，从而创建一个医学专家数据库，该数据库能够根据患者记录向医生提出治疗方案建议；还可以将该系统扩展为具有图像分析功能的系统，用于分析医学图像（如 X 射线和 CT 扫描）并进行预诊断。

使用大数据分析可以提高医疗服务质量。美国医疗保险和医疗补助服务中心（CMS）宣布了一项新的数据和信息计划，该计划由信息产品和数据分析办公室（OIPDA）管理。该计划将指导该机构从以"服务收费"为基础的付款人向"基于价值的护理购买者"转变，目的在于将支付与护理的质量和效率挂钩，而不是与纯粹的服务量挂钩。为了实现这一目标，需要通过分析从全国各地的医院和医生等服务提供者处收集的数据来衡量服务质量，以促进医疗服务行业标准的建立。大数据解决方案可以有效应对 CMS 和全球类似组织面临的挑战。

预测分析可以应用于分析患者档案，以确定易感染特定疾病的个体。可以进行预防性治疗以预防疾病或限制其严重程度，从而控制医疗成本。识别最需要服务的患者的能力对于提高治疗质量和节省资金具有重要意义。

5.7.2 零售

零售业建立在对消费者零售习惯的理解上。顶级零售商挖掘客户数据，并使用大数据技术来帮助他们做出有关营销、推销和供应链管理的决策。零售商使用更先进的方法来分析他们从多个销售渠道和互动中收集的数据。使用越来越细化的客户数据可以使零售商获得更详细的分析见解，进而提高其营销和推销工作的效率。更清晰的理解还将为零售商提供更准确的库存变动预测，从而改善供应链管理。

1. 营销

大数据分析可以通过多种方式应用于零售营销。其中一个应用是启用交叉销售，交叉销售使用客户的所有已知数据（包括客户的人口统计资料、购买历史和偏好）来增加平均购买量。在线零售商亚马逊就是一个很好的例子。凭借"猜你喜欢"的推荐引擎，亚马逊能够根据每个客户的浏览、购物习惯和其他在线个人资料，向他们推荐与其相关的产品。该公司称，其 30%的销售额是由其推荐引擎产生的。

另一个应用是分析顾客在店内的行为，以帮助零售商创造更有利的购物环境，增强顾客的购买倾向。这里有可能采用视频分析来分析顾客的店内流量模式和行为，以帮助改善商店布局、产品组合和货架定位。另外，大数据分析可以用来控制消费者的情绪。情绪分析利用了消费者在各种社交媒体平台上生成的数据流，以帮助微调各种营销决策。大数据分析帮助零售商创建其客户群的细分市场。随着客户数据的增加，分析工具的日益复杂使客户分析更加细化，帮助零售商提供更相关的（如果不是个性化的）产品和服务。

2. 推销

零售商必须考虑优化他们的产品组合。根据对当地人口统计资料、买家认知和购买习惯的分析来确定在商店里销售的合适的产品组合可以增加销售额。零售商可以使用越来越细化的定价、销售数据及分析工具来优化他们的定价策略。复杂的需求和弹性分析模型可用于检查历史销售数据，以深入了解最小存货单位（SKU）层面的产品定价。零售商也可以利用这些数据，并在考虑未来促销活动的情况下，找出可能推动销售的原因。

3. 供应链管理

库存管理在任何成功的零售策略中都是至关重要的。利用大数据分析提供的来自多个庞大数据集的详细信息，零售商可以清楚地了解他们的库存动向并改善其库存管理。执行良好的库存管理策略可以使零售商保持较低的库存水平，因为供应商可以更准确地响应消费者需求。零售商希望优化库存水平，以降低库存存储成本，并尽量减少因商品缺货而引起的销售损失。

5.7.3 教育

在教育界，学生在消费知识的同时也在创造信息。学生面临越来越高的课程要求，不再期望他们从难记的内容中照搬事实，而是要求他们对学习内容有深刻理解。同时，教育工作者也有责任为他们提供个性化的教学和指导。教育工作者面临的挑战是要有他们负责的每个学生的清晰档案。为每个学生创建个人档案将需要不同的信息集，这正是大数据分析的机会所在。

随着人们越来越重视学习的适应性、个性化和灵活性，有必要挖掘非结构化数据，如学生互动和学生生成的任何形式的内容。可以用学习分析工具来分析这些数据，如社交网络自适应教学实践工具（SNAPP）。SNAPP 是一种软件工具，可以使用户将讨论论坛帖子和回复产生的交互网络可视化。可视化为教师提供了机会，使其能够在课程进展的任何阶段快速识别用户的行为模式。此信息可快速识别任何在线学习活动的参与度和网络密度。从那里，可以识别出离线的和表现不佳的学生。网络可视化图的例子如图 5-1 所示。

大数据分析的其他应用可以以 Civitas Learning 为例进行说明。Civitas Learning 是一个数字教育平台，它使用预测分析来帮助指导教育决策。Civitas Learning 使用其合作机构提供的人口统计资料、行为和理论数据，并将其匿名化和合并。然后对数据进行分析，以确定趋势并提供见解,如根据学生档案和在确保良好教育效果方面最有效的教学方法,来确定最有益的课程和路径。这些见解将通过定制平台转化为对学生、教师和管理者的实时建议。

5.7.4 运输

1. 个性化的出行选项实时信息

智能手机的普及使个人可以使用信息工具来确定和提供各种出行选项,使通勤者能够依据时间和成本做出最佳决定。通过输入目的地,通勤者可以获得不同出行选项交通工具(包括公共汽车、火车或私家车)的预计到达时间。这需要将数据从安装在公共交通车辆上及沿路的传感器中输入到计算模型中,该模型可以预测一天中不同时间的交通方式和出行次数。该模型还可以用来计算其他因素,如在一天中的给定时间内特定道路上发生碰撞的可能性、天气状况,甚至预期的燃料消耗和成本。该模型可以预测火车和公共汽车到达的时间,并将这些公共交通方法与通勤者可能采用的不同乘车路线进行对比,并就如何以最有效的方式到达目的地提出建议。

2. 实时驾驶员辅助

对来自传感器网络的数据进行分析可以提供有助于驾驶员导航道路的信息,包括整合实时路况信息以启用主动线规划。司机会收到选择最不繁忙的道路的实时建议。随着算法复杂度的增加,车辆运行路线的重新规划将包括避开封闭道路和施工道路。这种方法可以以更平衡的方式优化道路的使用,从而在不需要昂贵的新交通网络的情况下,最大限度地缓解交通拥堵和减少总体旅行时间。

3. 公共运输系统调度

分析乘客登机数据、公共汽车和火车位置数据及一系列传感器网络数据,可以准确统计当前使用该系统的人数和每个站点的等待人数。可以根据实际需求,使用这些信息实现车辆的动态管理。

4. 预防性维护

对于拥有大量基础设施资产和定期维护计划的企业来说,预防性维护是一项主要的成本节约措施。定期维护计划是任何运输服务供应商确保运输车队(如出租车或公共汽车)可用的基本要求。然而,即使尽了最大努力进行维护,仍然会出现车辆故障导致运输服务中断的情况。在运输车队中,并不是所有车辆的维修工作都相同,与其他车辆相比,某些车辆需要更多的维修工作。很难确定需要更多维修工作的具体车辆。然而,通过应用数据分析,运输服务运营商将能够在停运前预测问题，并提高客户满意度。通过将车辆的日志数据捕获到分析平台上，可以执行算法来报告与车队中每辆车相关的问题和预测事件。与常规维护相比，预防性维护在确保运输车队的可用性方面更加有效。

5. 改善城市设计

交通系统的强化包括城市设计的改善。城市规划者可以通过收集和分析当地交通模式和人口密度数据，更准确地规划道路和公共交通建设，缓解交通拥堵。借助传感器和个人位置数据，城市开发商可以获得高峰和非高峰交通热点、公共交通使用模式和交通量信息。这些信息可以帮助城市规划者更准确地决定交通灯的位置和顺序，以及可能需要的停车位。

5.7.5 金融

大数据在金融业具有重要作用，特别是在应用复杂事件处理（CEP）进行欺诈检测方面。CEP 通过将看似无关的事件联系起来，使企业能够识别并预测机会和风险。CEP 通常通过实时聚合来自分布式系统的数据，并应用规则来识别模式和趋势，否则这些模式和趋势将不会被注意到。通过对这些庞大数据集的分析，可以更容易地检测到欺诈活动。例如，在中国餐馆支付账单几分钟后，从美国的 IP 地址在线购买法国火车票等不寻常的消费案例。然而，这只有在大型金融机构的 IT 系统能够从一系列来源自动采集和处理大量数据时才能实现，包括货币交易报告（CTR）、可疑活动报告（SARS）、可转让票据日志（NIL）及基于互联网的活动和事务。将概要文件更改的整个历史记录和交易记录包括在内是最理想的，以便在不同的集合和层次结构级别上，确定每个账户、客户、交易对手和法律实体的风险率。虽然处理能力和存储成本的限制使其在传统上不可能实现，但 HDFS 能够合并所有详细的数据点来计算此类风险的概况，并将结果发送到 CEP 引擎以建立风险模型的基础。数据库管理系统将捕获和存储不同来源大量低延迟数据及与 CEP 引擎集成的实时数据，以实现自动警报并触发业务流程，从而针对潜在的欺诈行为采取适当措施。

金融业的公司也应用大数据分析来了解客户行为，以增加客户亲密度并预计在适当的时间推出正确的产品。这涉及了解客户和竞争对手，及使用计算算法来了解世界。高性能分析有助于在正确的时间和地点，以正确的信息联系客户，因此银行可以获得并发展有益的客户群。许多公司（如科技型金融解决方案公司 British Pearl）利用非结构化数据寻找吸引和留住客户的最佳方式。由于获得客户的成本远高于维护客户的成本，美国的一家大型金融服务公司使用 1700 万客户和 1900 万笔日常交易的数据构成预警系统来检测客户脱离情况。某些互动和交易会触发警报，只要有迹象表明需要培养关系，一线员工就会立即与客户联系。金融公司不仅使用自己的数据集，还与金融业内外的合作伙伴合作，以获得对市场更全面、更准确的看法。通过获取与潜在客户及其在线状态有关的数据，可以更准确地为特定客户定制产品和服务，并且可以更有效地留住客户。

与各种数据源合作可以帮助金融机构确定衡量人们可能遇到的欺诈行为的模式和趋势，以降低银行的住房贷款风险。深入了解消费者的主要生活事件可以降低银行在客户整个生命周期中的风险敞口。为了评估信誉度并进行风险评估，银行可以应用高性能分析技术来检测各大型贷款组合中的细微异常。银行可以使用这些技术，将识别问题贷款所需的时间从100多小时缩短到几分钟，从而节省大量资金。

5.7.6 大数据隐私和安全

在大数据集的收集过程中，为了增强数据采样的随机性和减少偏差而采用的系统化数据收集方法并不明显。大数据集并不能自然消除数据偏差，所收集的数据仍然存在不完整和失真的可能，这反过来又可能导致出现错误的结论。以 Twitter 为例，通常会仔细检查 Twitter，以获取有关用户情绪的见解。使用 Twitter 作为数据源存在一个固有的问题，因为有 40%的 Twitter 活跃用户只是在接收信息，而并不发出信息。这表明推文可能仅来自某种类型的人（可能是那些在社交媒体中更善于表达和参与的人），而不是来自真正的随机样本。此外，Twitter 还通过其流式应用程序接口（API）向公众提供了其资料的样本。目前还不清楚该资料样本是如何获得的。

大型数据集的获取和大数据分析的使用显然涉及数据隐私和安全问题。因为信息很容易超越地理边界，所以数据安全和隐私保护工作变得更加困难。个人数据（如健康和财务数据）可以带来巨大利益，如帮助确定正确的医疗方法或适当的金融服务或产品。同样，个人购物和地理位置信息也可以通过向顾客提供更具相关性的产品和服务来为他们带来更好的购物体验。

企业可能已经尝试使用各种去识别化方法将数据与真实身份隔离，并允许在进行分析的同时解决隐私问题。然而，研究人员已经证明，匿名数据可以被重新识别并归于特定的人。这些研究成果通过破坏人们对匿名化的信任，扰乱了隐私政策的格局。去识别化已经成为众多商业模型的关键组成，尤其是在有针对性的营销和健康数据的背景下，这对政府和企业的影响显而易见。

Unit Six Big Data and Cloud Computing

6.1 Learning Goals

After learning this unit, you will be able to master the following knowledge:

1. In a broad sense, Cloud Computing refers to the mode of service delivery and use, through the network to obtain the required services in an on-demand and easy to develop way.

2. Compared with traditional data storage methods, Cloud Computing provides us with the most secure and reliable data storage center, because after users upload data to the cloud, they don't have to worry about data loss and virus invasion.

3. Cloud Computing can easily share data and applications between different devices.

4. Quantum Computing, a convergence between quantum physics and computers, represents this revolutionary form of computing where the data is represented by qubits.

6.2 Dialogue

Henry is talking with Mark about Cloud Computing in their dorm.

Henry: Excuse me, Mark. May I ask you some questions about Cloud Computing?

Mark: Sure. What can I do for you?

Henry: What is Cloud Computing?

Mark: Let me see. In my view, Cloud Computing is a new technology that provides flexible and massive online computing power. Apart from this, it's an Internet service that provides computing needs to computer users.

Henry: Can you give me an example?

Mark: Of course. An employee working during the day in California could use computing power in a network system located in an office that is closed for the evening. When the company uses the computing resources, they pay a fee based on the amount of computing time and other resources that they consume, much in the way that consumers pay utility companies, such as the electric company based on how much electricity they use.

Henry: I see. By the way, how did Cloud Computing start?

Mark: Cloud Computing began as large-scale Internet service providers, such as Google, Amazon, built out their infrastructure, massively scaled, horizontally distributed system resources, which is very beneficial to end users, because Cloud Computing means there are no hardware acquisition costs, no software licenses or upgrades to manage, no new employees or consultants to hire, no facilities to lease, no capital costs of any kind.

Henry: That's fantastic.

Mark: Yes, indeed. In your life, you must have many account numbers like QQ, Baidu Netdisk. With these account numbers, you can send the music, the pictures, the movies you like or the documents you need to your cloud space so that you can get access to your resources online conveniently, without occupying the storage of your smartphone or your computer.

Henry: Wow, how nice! Cloud Computing is changing our life definitely!

6.3 Reading

6.3.1 What is Cloud Computing?

When it comes to Cloud Computing, you might ask, "what exactly is it?" Like other conceptual computing, it is a computing method of the Internet. On this basis, it realizes the sharing of hardware and software resources and information. The network resources it provides are usually virtualized and have the characteristics of dynamic and easy expansion. This network application mode is mainly based on the increase, use and delivery of Internet related services, which was first proposed by Google.

In fact, "cloud" is a metaphor for the Internet. People have visualized the way of data calculation. In the past, "cloud" was used to represent telecommunication networks; Later, "cloud" was also used as an abstract concept to represent the Internet and underlying infrastructure. In a narrow sense, Cloud Computing is the delivery and use mode of IT infrastructure, which refers to obtaining the required resources in an on-demand and easy to expand way through the network; In a broad sense, Cloud Computing refers to the mode of service delivery and use, through the network to obtain the required services in an on-demand and easy to develop way. This can be not only IT, software, Internet related services, but also applications in other fields. Moreover, this means that computing can also be used as a commodity to circulate through the Internet.

In the United States, NIST defines Cloud Computing as: Cloud Computing is a pay as you use model, which provides available and convenient on-demand network access to the configurable computing resource sharing pool (resources include networks, servers, storages, application softwares and services). These resources can be provided quickly with little management or interaction with service providers.

Once the concept of "Cloud Computing" is put forward, it is widely used in the production environment. "Alibaba cloud" in China and Xen System of CloudValley, as well as Intel and IBM, which are already very mature abroad, are expanding the application service scope of "Cloud Computing" day by day. It can be said that in the era of Big Data, the impact of Cloud Computing in the future is immeasurable.

"Cloud Computing" is often confused with the concepts of grid computing, utility computing and autonomic computing. Next, we will make a basic explanation and distinction for the concepts of these calculation methods.

Grid computing: a kind of distributed computing, a super virtual computer composed of a group of loosely coupled computers, which is often used to perform some large tasks.

Utility computing: a way to package and charge resources, such as calculating and storing the cost separately, like the traditional power and other public facilities.

Autonomic computing: a computer system with self-management functions.

In fact, many Cloud Computing technology deployments rely on computer clusters, which also absorb the characteristics of autonomic computing and utility computing. However, it should be noted that Cloud Computing is quite different from grid computing in terms of composition, architecture, purpose and working mode.

6.3.2 Safe and Reliable Storage Data

In the past, people's traditional way of data storage is to store in the hard disk, many users suffer from virus, Trojan horse attacks, or due to some negligent operation resulting in data loss. Compared with traditional data storage methods, Cloud Computing provides us with the most secure and reliable data storage center, because after users upload data to the cloud, they don't have to worry about data loss and virus invasion. If the computer breaks down or even completely destroys, you can still recover the data through cloud on another computer.

But this kind of storage method is not accepted by everyone. Many people think that data storage should be the same as bank card saving, and only when it is stored in a place they can see, can they feel secure, so they will think that their computers are the safest. But what is the fact? As you can see, your computer will be damaged or attacked, and some lawbreakers will steal your computer data by various means.

If the document is saved on a Web service similar to Google Docs, such as your own private photos and videos, you will no longer have to worry about the unexpected trouble caused by the loss or damage of the data. At the other end of the cloud, a professional team will manage your information and save data for you. Moreover, you don't need to worry about data disclosure. Because cloud storage has strict permission management, you can safely share data with the people you specify.

6.3.3 Low Requirements of Cloud Computing for Clients

Cloud Computing has the lowest requirements for the devices of the user end and is easy to operate, which is not available in other storage methods. For example, we often need to update various application softwares. Sometimes, in order to use the latest software version or operating system, we have to spend a lot of time upgrading computer hardware. The most troublesome thing is that a friend sent us a document in a certain format, and we had to download and install an application software according to the requirements of this document. For this reason, the computer is equipped with a large number of applications that may only be used once, which greatly reduces the running speed of the computer. In addition, in order to prevent viruses that may attack at any time during downloading, anti-virus software and firewall must be loaded.

These trivial procedures add up as if they are in constant trouble. Sometimes it takes tens of times to read a document with hundreds of words. And you happen to be a computer novice, this experience is definitely a dream for you!

At this time, the advantages of Cloud Computing are reflected. It will bring you a new and simple experience. First of all, you only need to have a computer that can access the Internet; Secondly, there is a browser you like on your computer. Next, you just need to type the URL in the browser, and then you can enjoy the unlimited fun brought by Cloud Computing.

You can edit the documents stored on the other end of the cloud directly in the browser, and you can share information with friends at any time, and you no longer need to worry about whether the software in the computer is the latest version, and no longer worry about the annoying virus. At the other end of the cloud, there is professional IT personnel to help you maintain your hardware, install and upgrade software, protect you from viruses and all kinds of network attacks, and do everything you have done on your personal computer before. It's like having a loyal and competent housekeeper. You just need to enjoy comfort and safety at home quietly. He will take care of all the trifles for you.

6.3.4 Easy Data Sharing

Cloud Computing can easily share data and applications between different devices. This greatly reduces the trouble of data transfer. For example, your mobile phone stores hundreds of contact information, but when you buy a new mobile phone, you have to transfer the number on the old mobile phone to the new one, which needs to be synchronized. The same is true for home computers and office computers, which need to be synchronized frequently to ensure that no information is missed.

Because different devices have many methods of data synchronization, and the operation is more complex, it will cost a lot of time to realize the desire of "the latest contact list" among these different devices. And Cloud Computing can make everything simple.

In the network application mode of Cloud Computing, there is only one copy of data, which is saved at the other end of the "cloud". All your electronic devices can access and use the same data at the same time only by connecting to the Internet. In other words, the data is always up to date.

6.3.5　It Could be Infinite

Cloud Computing provides us with unlimited possibilities to use the network, which can be achieved by almost any imagination. For example, you plan to drive with your family when it's sunny. At this time, you want to check the traffic in your location. You only need to connect your mobile phone to the network, and you can quickly and clearly see the satellite map of your current location, and master the traffic conditions in a timely and convenient manner from above, You can also quickly check your preset driving route on the way out, and communicate with your friends on the Internet in real-time to find the best restaurants, hotels and beautiful scenery recommended by others. You can also use your fingers to reserve hotels at the destination. All of this is so wonderful. Of course, we should not forget to share when enjoying the experience. You can clip the photos or videos you just took and share them with your friends and relatives who are paying close attention to you from afar.

If there is no Cloud Computing, just using the client applications on personal computers or mobile phones, these infinite possibilities will not be experienced. Not only that, in terms of storage capacity, personal computers or other electronic devices are clumsy and "stingy", because the storage space of hard disk is always limited and can not provide us with unlimited computing power, but "cloud" can easily achieve this. At the other end of the cloud, there is a huge cluster of tens of thousands or more servers. We can hardly imagine how much data storage and computing can be undertaken at the other end of the cloud, because it is almost infinite. Cloud Computing well embodies the spirit of the Internet — freedom, equality and sharing. In the era of Big Data, Cloud Computing has shown infinite vitality. In the future, it will also change and affect our work and life, and greatly improve our quality of life.

Whether you are an ordinary network user, an employee of an enterprise, an IT manager, or a novice who just touches the computer, as long as you use Cloud Computing, you can experience this obvious change in a person.

6.3.6 Quantum Computing

Modern computing may be considered powerful by today's standards. However, to increase the computing power of conventional computers, it is necessary to pack more transistors (the basic components of a computer) within a single computer. This is getting increasingly difficult. The maximum number of transistors that can possibly be packed will soon be reached and computers will have to be revolutionized to meet the computing demands of the future.

Quantum Computing, a convergence between quantum physics and computers, represents this revolutionary form of computing where the data is represented by qubits. Unlike conventional computers which contain many small transistors that can either be turned on or off to represent 0 or 1 to represent data, each qubit in Quantum Computing can be in the state of 0 or 1, or a mixed state where it represents 0 and 1 at the same time. This is a property known as superposition in quantum mechanics and it provides quantum computers the ability to compute at a speed faster than conventional computers. For certain problems like searching databases or factoring very large numbers (the basis of today's encryption techniques), quantum computer could produce an answer in days whereas the fastest conventional computer would take longer than 13.7 billion years of computation.

In Quantum Computing, data value held in a qubit has a strong correlation with other qubits even when they are physically separated. This phenomenon, known as entanglement, allows scientists to dictate the value of one qubit just by knowing the state of another qubit. Qubits are highly susceptible to the effects of noise from the external environment. In order to ensure accuracy in Quantum Computing, qubits must be linked and placed in an enclosed quantum environment, shielding them from noise. Areas that require huge computational power, e.g., security and digital image processing, will witness Quantum Computing transforming the speed at which these processes are carried out.

To date, Quantum Computing has not been demonstrated in a verifiable way as the technology is still in the early stage of research. However, Quantum Computing continues to attract significant funding and research is being carried out on both the hardware and algorithm design. Some significant advances made during recent years may pave the way for the development of a quantum computer.

Google has been using a Quantum Computing device created by D-Wave since 2009 to research on a highly efficient way to search for images based on an improved quantum algorithm discovered by researchers at MIT.

IBM researchers built on a technique developed by Robert J. Schoelkopf, a physics professor at Yale, derived a qubit in 2012. This helped to lengthen the time in which error correction algorithms can detect and fix mistakes. Otherwise, generating reliable results from Quantum Computing is impossible as the error rate is too high.

6.4 New Words & Phrases & Sentences

6.4.1 New Words

1. conceptual — *adj.* 概念（上）的；观念（上）的
2. hardware — *n.* 硬件；（家庭及园艺用）工具，设备，五金制品；硬件设备；机器；车辆
3. dynamic — *n.* （人或事物）相互作用的方式，动态；力学；动力学；动力
   adj. 充满活力的；精力充沛的；个性强的；动态的
4. clip — *n.* 夹子，回形针；修剪；电影片段；（用手）猛击，抽打；子弹夹
   v. 夹住；剪（掉），修剪；碰撞
5. demonstrate — *vt.* 示范，证明，论证
   vi. 示威
6. qubit — *n.* 量子位

7. abstract   *adj.* 抽象的

　　　　　　　　n.（文献等的）摘要，概要

　　　　　　　　v. 把……抽象出；提取；抽取；分离；写出（书等的）摘要

8. narrow   *adj.* 狭窄的；勉强的；狭隘的；目光短浅的

9. circulate   *v.*（液体或气体）环流，循环；传播；流传；散布；传送；传递；传阅

10. interaction   *n.* 相互影响（作用，制约，配合）；交互作用（影响）；交相感应；干扰

11. shield   *n.* 盾（牌）；保护人；保护物；掩护物；屏障；（保护机器和操作者的）护罩，防护屏

　　　　　　　　v. 保护某人或某物（免遭危险、伤害）；给……加防护罩

12. immeasurable   *adj.* 不可估量的；无限的；无穷的

13. flexible   *adj.* 能适应新情况的；灵活的；可变动的；柔韧的；可弯曲的；有弹性的

14. equality   *n.* 平等；均等；相等

15. quantum   *n.* 量子

6.4.2　Phrases

1. network application mode   网络应用模式
2. large-scale Internet service   大型互联网服务
3. telecommunication network   电信网络
4. In a narrow sense   从狭义上讲
5. In a broad sense   从广义上讲

6.	horizontally distributed	水平分布的
9.	on-demand network access	按需网络接入
10.	be confused with	与……混淆
11.	grid computing	网格计算
12.	hardware acquisition cost	硬件购置成本
13.	utility computing	效用计算
14.	autonomic computing	自主计算
15.	Safe and reliable storage data	数据存储安全可靠

6.4.3 Sentences

1. Like other conceptual computing, it is a computing method of the Internet. On this basis, it realizes the sharing of hardware and software resources and information. 与其他概念计算一样，它是互联网的一种计算方式。在此基础上，它还实现了软硬件资源和信息的共享。

2. If the computer breaks down or even completely destroys, you can still recover the data through cloud on another computer. 就算计算机出现故障甚至完全毁坏，你也可以在另一台计算机上通过云复原数据。

3. As you can see, your computer will be damaged or attacked, and some lawbreakers will steal your computer data by various means. 正如你所看到的，你的计算机会被破坏或攻击，有些不法分子也会用各种手段窃取你的计算机数据。

4. At the other end of the cloud, a professional team will manage your information and save data for you. 在云的另一端，有专业的团队会管理你的信息，为你保存数据。

5. Moreover, you don't need to worry about data disclosure. Because cloud storage has strict permission management, you can safely share data with the people you specify. 而且，你不需要担心数据泄露的问题，因为云存储有严格的权限管理，你可以放心地与你指定的人共享数据。

6. Cloud Computing has the lowest requirements for the devices of the user end and is easy to operate, which is not available in other storage methods. 云计算对用户端设备的要求最低，操作起来方便简单，这是其他存储方式无法提供的。

7. The most troublesome thing is that a friend sent us a document in a certain format, and we

had to download and install an application according to the requirements of this document. 最麻烦的是，有朋友发来了某种格式的文档，我们就不得不根据这个文档的需求来下载并安装某个应用软件。

8. In addition, in order to prevent viruses that may attack at any time during downloading, anti-virus software and firewall must be loaded. 而且，为了阻止在下载时随时可能侵袭而来的病毒，必须加载杀毒软件和防火墙。

9. At this time, the advantages of Cloud Computing are reflected. It will bring you a new and simple experience. 这时候，云计算的优势就体现了出来，它将带给你全新的简洁体验。

10. Cloud Computing provides us with unlimited possibilities to use the network, which can be achieved by almost any imagination. 云计算为我们使用网络提供了无限的可能，几乎可以实现任何想象。

6.5 Exercises

【Ex. 1】Content Questions.

1. What is Cloud Computing?

2. What are easily confused with Cloud Computing?

3. What are the characteristics of Cloud Computing?

4. What are the requirements for computers to use Cloud Computing?

【Ex. 2】补充空白部分的单词。

1. As you can see, your computer will be _____ or _____, and some lawbreakers will _____ your computer data by various means.

2. In a narrow sense, Cloud Computing is the delivery and use mode of IT _____, which refers to obtaining the _____ _____ in an on-demand and easy to expand way through the network.

3. This can be not only IT, software, Internet _____ services, but also _____ in other fields.

4. Cloud Computing is a pay as you use model, which provides _____ and _____ on-

demand network access to the _____ computing resource sharing pool (resources include networks, servers, storages, application softwares and services).

5. These resources can be _____ quickly with little _____ or _____ with service providers.

6. In the past, people's traditional way of data _____ is to store in the hard disk, many users suffer from virus, Trojan horse attacks, or _____ to some negligent _____ resulting in data loss.

7. If the computer _____ _____ or even completely _____, you can still recover the data through cloud on another computer.

8. At the other end of the cloud, a professional team will _____ your information and _____ data for you.

6.6 Translations for Dialogue

亨利正在宿舍和马克谈论云计算。

亨利：打扰一下，马克。我可以问你一些关于云计算的问题吗？

马克：当然可以。我能为你做些什么？

亨利：什么是云计算呢？

马克：让我想想。在我看来，云计算是一种新的技术，可以提供灵活和庞大的在线计算能力。除此之外，它还是一种为计算机用户提供计算需求的互联网服务。

亨利：你能举个例子吗？

马克：当然可以。在加州，员工白天可以使用办公室中网络系统的计算能力，该系统晚上将会关闭。当公司使用计算资源时，他们会根据计算时间和消耗的其他资源的数量支付费用，这与消费者根据使用电量向公用事业公司（如电力公司）支付费用的方式非常相似。

亨利：我明白了。顺便问一下，云计算是如何开始的？

马克：云计算始于大型互联网服务供应商，如谷歌、亚马逊，他们建立了基础设施和大规模、水平分布的系统资源，这对于终端用户来说非常有利，因为云计算意味着没有硬件购置成本，没有软件许可证或升级管理，没有新员工或顾问可以雇用，没有设施可以租赁，没有任何形式的资本成本。

亨利：这真是太棒了。

马克：是的，的确如此。在你的生活中，一定有许多账号，如QQ、百度网盘。你可以使用这些账号将音乐、图片、喜欢的电影或所需的文档发送到云空间，以便方便地访问你的在线资源，而不占用智能手机或计算机的存储空间。

亨利：哇，真棒！云计算确实正在改变我们的生活！

6.7 Translations for Reading

6.7.1 什么是云计算？

提到云计算，你可能会问，"它到底是什么？"与其他概念计算一样，它是互联网的一种计算方式。在此基础上，它还实现了软硬件资源和信息的共享。它提供的网络资源通常是虚拟化的，具有动态和易扩展的特点。这种网络应用模式主要基于最早由谷歌提出的互联网相关服务的增加、使用和交付。

事实上，"云"是互联网的一种隐喻的说法，人们使数据的计算方式变得可视化。过去，"云"被用来表示电信网络；后来，"云"也被用来作为表示互联网和底层基础设施的抽象概念。从狭义上讲，云计算是IT基础设施的交付和使用模式，指通过网络以按需且易扩展的方式获得所需资源；从广义上讲，云计算指通过网络以按需且易扩展的方式获得所需服务的服务交付和使用模式，既可以是与IT、软件、互联网相关的服务，也可以是其他领域的应用。此外，这还意味着计算也可以作为一种商品在互联网上流通。

在美国，NIST将云计算的定义为：云计算是一种按使用量付费的模型，这种模型提供可用的、便捷的按需网络接入，可以进入可配置的计算资源共享池（资源包括网络、服务器、存储、应用软件、服务），只需投入很少的管理工作或与服务供应商进行很少的交互就能快速获得这些资源。

"云计算"的概念一经提出，就被大量运用到生产环境中。国内的"阿里云"与云谷公司的Xen System，以及在国外已经非常成熟的Intel和IBM，都在日益扩大"云计算"的应用服务范围。可以说在大数据时代，云计算未来的影响不可估量。

"云计算"的概念常常与网格计算、效用计算、自主计算的概念混淆。下面我们对这几种计算方式的概念进行基本的解释和区分。

网格计算：分布式计算的一种，是由一群松散耦合的计算机组成的一个超级虚拟计算机，常用来执行一些大型任务。

效用计算：一种资源打包和收费方式，如像传统的电力和其他公共设施一样分别计算和存储成本。

自主计算：具有自我管理功能的计算机系统。

事实上，许多云计算技术的部署都依赖于计算机集群，而计算机集群也吸收了自主计算和效用计算的特点。然而，需要指出的是，云计算与网格计算在组成、体系结构、目的、工作方式方面大相径庭。

6.7.2 存储数据安全可靠

过去，人们采用的传统数据存储方式是硬盘存储，很多用户遭受病毒、特洛伊木马的攻击，或者由于一些疏忽操作导致数据丢失。与传统的数据存储方式相比，云计算为我们提供了最安全可靠的数据存储中心，因为用户把数据上传到云以后，不用再担心数据丢失和病毒入侵问题。就算计算机出现故障甚至完全毁坏，你也可以在另一台计算机上通过云复原数据。

但这种存储方式并不被所有人接受，很多人觉得数据存储应该和银行卡储蓄一样，只有保存在自己看得见的地方心里才踏实，所以他们会以为自己的计算机最安全。但事实是什么？正如你所看到的，你的计算机会被破坏或攻击，有些不法分子也会用各种手段窃取你的计算机数据。

如果将文档保存在类似 Google Docs 的网络服务上（如自己的私密照片、视频等），则无须担心数据的丢失或损坏给自己带来意想不到的麻烦。在云的另一端，有专业的团队会管理你的信息，为你保存数据。而且，你不需要担心数据泄露的问题，因为云存储有严格的权限管理，你可以放心地与你指定的人共享数据。

6.7.3 云计算对客户端的要求低

云计算对用户端设备的要求最低，操作起来方便简单，这是其他存储方式无法提供的。例如，我们经常需要更新各种应用软件，有时候为了能够使用最新版本的软件或操作系统，我们不得不花费大量时间升级计算机硬件。最麻烦的是，有朋友发来了某种格式的文档，我们就不得不根据这个文档的需求来下载并安装某个应用软件。为此，计算机上装了一大堆可能只会使用一次的应用软件，大大降低了计算机的运行速度。而且，为了阻止在下载时随时可能侵袭而来的病毒，必须加载杀毒软件和防火墙。

这些琐碎的程序加在一起，就像遇到了接踵不断的麻烦。有时候为了看一个几百字的文档需要花费几十倍的时间。若你恰巧是一个计算机新手，这种体验对你来说绝对是一场噩梦！

这时候，云计算的优势就体现了出来，它将带给你全新的简洁体验。首先，你只需要有一台可以上网的计算机。其次，你的计算机上有你喜欢的浏览器。接下来，你只需要在浏览器中输入 URL，即可尽情享受云计算带给你的无限乐趣了。

你可以在浏览器中直接编辑存储在云的另一端的文档，也可以随时与朋友分享信息，不再需要担心计算机中的软件是否是最新版本，也不再需要为恼人的病毒而发愁。在云的另一端，有专业的 IT 人员帮你维护硬件，帮你安装和升级软件，帮你防范病毒和各类网络攻击，帮你做你以前在个人计算机上所做的一切。这就如同请了一个忠实能干的管家。你只需要在家里安静地享受舒适与安全，他会替你打理一切琐事。

6.7.4 轻松共享数据

云计算可以轻松地实现不同设备间的数据与应用共享，大大减少了数据传输的麻烦。例如，你的手机里存储了几百个联系人的信息，可是当你买了新手机之后，不得不把旧手机上的号码转移到新手机上，这就需要同步。家里的计算机和办公室的计算机也是如此，需要经常进行同步才能保证信息不遗漏。

由于不同设备有许多数据同步方法，操作起来更加复杂，想要在这么多设备之间实现"最新联系人列表"的愿望，将花费大量时间。而云计算能够让一切变得简单。

在云计算的网络应用模式中，数据只有一份，并保存在云的另一端。你的所有电子设备只需要连接互联网，就可以同时访问和使用同一份数据。也就是说，数据永远是最新的。

6.7.5 无限的可能性

云计算为我们使用网络提供了无限的可能，几乎可以实现任何想象。例如，你打算在天气晴朗的时候和家人驾车出行，这时想要查看一下自己所在位置的交通，你只需要将手机连入网络，就可以快速、清晰地查看当前位置的卫星地图，及时、便捷地掌握交通状况。你还可以在出行的途中快速查询自己预设的行车路线，与网络上与好友实时交流以找到他人推荐的附近最好的餐馆、酒店和美丽的风景。你也可以自行预订目的地的酒店。这一切都是如此美妙，当然了，我们在享受这些经历的时候不要忘记分享，你可以对自己刚刚拍摄的照片或视频进行剪辑，分享给正在远方密切关注你的亲朋好友。

如果没有云计算，仅使用个人计算机或手机上的客户端应用程序，则不会经历这些。不但如此，就存储量来讲，个人计算机或其他电子设备显得又笨拙又"小气"，因为硬盘的存储空间总是有限的，无法为我们提供无限的计算能力，但"云"可以轻松实现这点。在云的另一端，有由成千上万台甚至更多服务器组成的庞大集群，我们几乎无法想象云的另一端到底能承载多少数据存储和计算，因为那几乎是无限的。云计算很好地体现了互联网的精神实质——自由、平等和分享。在大数据时代，云计算已经展现出了无穷的生命力，未来，它将更多地改变和影响我们的工作和生活，大大提高我们的生活质量。

无论你是普通网络用户、企业的员工、IT管理者，还是刚刚接触计算机的新手，只要你使用了云计算，就能亲身体验到这种显而易见的改变。

6.7.6　量子计算

按照如今的标准，可能会认为现代计算功能强大。然而，为了提高传统计算机的计算能力，有必要在一台计算机内装更多的晶体管（计算机的基本部件），但这变得越来越困难，很快将达到可以封装的晶体管的最大数量，计算机将不得不进行革命性的改变以满足未来的计算需求。

量子计算是量子物理和计算机的融合，它代表了一种用量子位表示数据的革命性计算形式。传统计算机包含许多小晶体管，这些晶体管可以打开或关闭，即用0或1来表示数据。与传统计算机不同，量子计算中的每个量子位可以处于0或1状态，也可以同时表示0和1，处于混合状态。这是量子力学中的一种叠加性质，它使量子计算机能够以比传统计算机快得多的速度进行计算。对于某些问题，如搜索数据库或分解大的数字（当今加密技术的基础），量子计算机可以在几天内给出答案，而最快的传统计算机则需要137亿年以上的计算时间。

在量子计算中，即使在物理隔离的情况下，量子位中的数据值也与其他量子位有很强的相关性。这种被称为纠缠的现象使科学家能够通过知道另一个量子位的状态来决定一个量子位的值。量子位极易受外界噪声的影响。为了保证量子计算的精确性，必须将量子位连接起来并放置在一个封闭的量子环境中，以使其免受噪声干扰。需要巨大计算能力的领域，如安全和数字图像处理，将见证量子计算改变这些过程的执行速度。

迄今为止，量子计算还没有以一种可验证的方式得到证明，因为这项技术仍处于研究的早期阶段。但是，量子计算持续吸引大量资金，并且正在进行硬件和算法设计方面的研究。近年来取得的一些重大进展可能会为量子计算机的发展铺平道路。

2009 年以来，谷歌一直使用 D-Wave 开发的量子计算设备，以麻省理工学院研究人员发现的一种改进量子算法为基础，研究一种高效的图像搜索方法。

IBM 研究人员基于耶鲁大学物理教授 Robert J. Schoelkopf 开发的一项技术，于 2012 年推导出了一个量子位。这有助于延长纠错算法检测和修复错误的时间。否则，错误率太高将导致无法从量子计算中生成可靠的结果。

07 Unit Seven   Big Data and Blockchain

7.1 Learning Goals

After learning this unit, you will be able to master the following knowledge:

1. The data record of blockchain is unchangeable and permanent.

2. The distributed storage of blockchain is to store all records in multiple accounting nodes of the whole network. The damage or loss of a single node will not affect other nodes, and the data error or tampering of a single node is more unlikely to have any destructive impact on the overall data.

7.2 Dialogue

Sophie is asking Professor Jackson about Bitcoin in his office.

Sophie: What is Bitcoin?

Prof Jackson: Bitcoin is a kind of digital currency. You can buy it with dollars or euros, just like you can trade any other currency. You store it in an online "wallet". And with that wallet, you can spend Bitcoin online and in the physical world for goods and services. And, of course, Bitcoin has a valuation, which you may have heard about Bitcoin's price has fluctuated up and down.

Sophie: What's different about Bitcoin?

Prof Jackson: Usually, if you pay for something on the Internet, you use a credit or debit card. That card is connected to information about you, such as your name and billing address. You can use Bitcoin in the same way, but unlike with a credit card, the transactions you make with the currency are completely anonymous. They can't be used to identify you personally.

Sophie: So you can use Bitcoin to protect your privacy. Is that why the WannaCry attackers picked it as a form of payment?

Prof Jackson: Possibly. Bitcoin has certainly gained prominence in the news media as a technology that can facilitate crime. But even though the identities of people in a Bitcoin transaction may be hidden, the public ledger has increasingly helped law enforcement trace the movements of bitcoins from place to place.

Sophie: I got it. Thanks.

Prof Jackson: You're welcome.

7.3 Reading

As a new information technology, blockchain uses time stamp and digital password technology to record transaction records in the data block composed of time series, and uses consensus mechanism to store data in the distributed database, so as to generate the unique data record that is permanently saved and irreversibly tampered, and achieve the purpose of realizing credible transaction without relying on any central organization.

The data record of blockchain is open and transparent. A more vivid example is that in the era without network and telephone, people's remote information exchange is mainly through the post office. The difference between blockchain ledger database and traditional database is just like the difference between postcard and ordinary mail in that era: the content of the postcard may be seen by many people in or around the post office, and the receiver and sender want to deny it, but they can't deny it. But ordinary letters are different. Outsiders don't know what they are about. At the same time, the data record of blockchain is unchangeable and permanent. This is like multiple copies of invoices or receipts. Modifying or destroying a single document cannot change the data records of other documents.

Around 17 p.m. on May 27, 2015, many netizens across the country reflected that Alipay could not log in. At about 18:00 p.m., Alipay released a message on Weibo, admitting that there was a failure in the use of Alipay due to the excavation of an optical cable in Xiaoshan District, Hangzhou city. This event typically reflects the huge risks of the centralized database. Whether the database itself is damaged (the computer room is in trouble, the database data is damaged or tampered), or the communication between the central database and the external terminal is in trouble, the centralized database will not work. The distributed storage of blockchain is to store all records in multiple accounting nodes of the whole network. The damage or loss of a single node will not affect other nodes, and the data error or tampering of a single node is more unlikely to have any destructive impact on the overall data. The data stored in different nodes are strictly protected by cryptography technology. Even if the relevant information is obtained, the real content of data information cannot be peeped without legal authorization.

Through consensus mechanism, blockchain creatively solves the problem of information synchronization of all accounting nodes in the whole network. It can effectively get rid of the influence of some problem nodes and complete the correct accounting update. The PoW consensus mechanism represented by Bitcoin can also encourage network nodes to participate in bookkeeping by providing effective feedback, attracting computing power from all over the world to serve as the bookkeeper for Bitcoin, thus effectively maintaining the operation and development of the whole Bitcoin blockchain network.

7.4 New Words & Phrases & Sentences

7.4.1 New Words

1. destructive　　　　　　　　　*adj.* 引起破坏（或毁灭）的；破坏（或毁灭）性的
2. Bitcoin　　　　　　　　　　*n.* 比特币

3. irreversibly   *adv.* 不可逆地

4. tamper   *v.* 干预；篡改

5. transparent   *adj.* 透明的；清澈的；易识破的；易看穿的；显而易见的；易懂的

6. prominence   *n.* 重要；突出；卓越；出名

7. centralize   *v.* 集权控制；实行集中

8. permanently   *adv.* 永久地；永远；长期；一直

9. cryptography   *n.* 密码学；密码术

10. consensus   *n.* 一致的意见；共识

11. participate   *v.* 参加；参与

12. netizen   *n.* 网民

13. synchronization   *n.* 同时；同时性；同步；同期

14. mechanism   *n.* 机械装置；机件；方法；机制；（生物体内的）机制，构造

15. peep   *n.* 偷偷一瞥；说话；出声音；啾啾声；嘟嘟声

7.4.2 Phrases

1. without legal authorization   没有合法授权
2. consensus mechanism   共识机制
3. distributed database   分布式数据库
4. digital currency   数字货币
5. destructive impact   破坏性影响
6. fluctuate up and down   上下波动
7. cryptography technology   加密技术
8. be protected by   受……保护

9. credible transaction　　　　　　　　　可信交易

10. blockchain ledger database　　　　　　区块链分类账数据库

11. the whole Bitcoin blockchain network　　整个比特币区块链网络

12. facilitate crime　　　　　　　　　　　助长犯罪

13. time stamp　　　　　　　　　　　　　时间戳

7.4.3　Sentences

1. The data record of blockchain is open and transparent. A more vivid example is that in the era without network and telephone, people's remote information exchange is mainly through the post office. 区块链的数据记录是公开透明的，一个比较形象的例子就是在没有网络和电话的年代，人们的远程信息交流方式主要是通过邮局寄信。

2. The content of the postcard may be seen by many people in or around the post office, and the receiver and sender want to deny it, but they can't deny it. 明信片的内容可能会被邮局的人或周围的许多人看到，收信人和发信人想否认也否认不了。

3. At the same time, the data record of blockchain is unchangeable and permanent. 同时，区块链的数据记录不可篡改且永久不变。

4. This is like multiple copies of invoices or receipts. Modifying or destroying a single document cannot change the data records of other documents. 就像发票或收据的多份副本，修改或销毁单张单据无法改变其他单据的数据记录。

5. This event typically reflects the huge risks of the centralized database. 这个事件非常典型地反映了集中式数据库存在的巨大风险。

6. The data stored in different nodes are strictly protected by cryptography technology. Even if the relevant information is obtained, the real content of data information cannot be peeped without legal authorization. 存储在不同节点的数据信息都受加密技术的严格保护，即使获得了相关信息，没有合法授权也无法偷窥到数据信息的真实内容。

7. Through consensus mechanism, blockchain creatively solves the problem of information synchronization of all accounting nodes in the whole network. It can effectively get rid of the influence of some problem nodes and complete the correct accounting update. 区块链通过共识机制创造性地解决了全网各记账节点的信息同步问题，可以有效摆脱某些问题节点的影响，完

成正确的记账更新。

7.5　Exercises

【Ex. 1】Content Questions.

1. What is blockchain?

2. What is the difference between the blockchain ledger database and traditional database?

3. How does the blockchain complete the bookkeeping update?

4. How blockchain is stored?

【Ex. 2】补充空白部分的单词。

1. A more vivid example is that in the era without _____ and _____, people's remote information _____ is mainly through the post office.

2. The _____ between _____ ledger database and traditional database is just like the difference between postcard and _____ mail in that era.

3. At the same time, the data record of blockchain is _____ and _____.

4. This event typically _____ the huge risks of the centralized _____.

5. The _____ _____ of blockchain is to store all records in _____ accounting nodes of the whole network.

6. The _____ or loss of a single node will not affect other nodes, and the data error or _____ of a single node is more unlikely to have any _____ impact on the overall data.

7. The data stored in different nodes are strictly _____ by _____ technology.

8. Through consensus mechanism, blockchain _____ solves the problem of information _____ of all accounting nodes in the whole network.

7.6 Translations for Dialogue

索菲正在杰克逊教授的办公室问他咨询比特币的事。

索菲：什么是比特币？

杰克逊教授：比特币是一种数字货币。你可以用美元或欧元购买它，就像你可以交易任何其他货币一样。你可以把它储存在一个在线的"钱包"里。有了这个钱包，你可以在网上和现实世界中花费比特币购买商品和服务。当然，比特币也有估值，你可能听说过比特币的价格上下波动。

索菲：比特币有什么不同呢？

杰克逊教授：在通常情况下，如果在网上付款，你会使用信用卡或借记卡。这张卡与你的信息有关，如你的名字和账单地址。你可以用同样的方式使用比特币，但与信用卡不同的是，你用这种货币做的交易完全是匿名的。它们不能用来识别你的个人身份。

索菲：因此你可以用比特币来保护你的隐私。这就是永恒之蓝攻击者选择它作为一种支付方式的原因？

杰克逊教授：可能吧。比特币作为一种可以助长犯罪的技术，无疑在新闻媒体中取得了突出的地位。不过，尽管在比特币交易中，人的身份可能被隐藏，但公共账本越来越有助于执法部门追踪比特币在各地的流动情况。

索菲：我明白了，谢谢。

杰克逊教授：不客气。

7.7 Translations for Reading

作为一种新兴的信息技术，区块链使用时间戳和数字密码技术，把交易记录记载在按时间序列组成的数据区块中，并使用共识机制把数据存储到分布式数据库内，从而生成了永久保存、不可逆向篡改的唯一数据记录，达到不依靠任何中心机构实现可信交易的目的。

区块链的数据记录是公开透明的，一个比较形象的例子就是在没有网络和电话的年代，人们的远程信息交流方式主要是通过邮局寄信。区块链分类账数据库与传统数据库的差别就像那个年代的明信片与平信的差别：明信片的内容可能会被邮局的人或周围的许多人看到，收信人和发信人想否认也否认不了；而平信则不同，外人并不知道信里写的内容是什么。同时，区块链的数据记录不可篡改且永久不变。就像发票或收据的多份副本，修改或销毁单张单据无法改变其他单据的数据记录。

2015年5月27日17时左右，全国多地网友反映支付宝无法登录。18时左右，支付宝通过微博发布消息，承认支付宝使用出现故障，原因是杭州市萧山区某地光缆被挖断。这个事件非常典型地反映了集中式数据库存在的巨大风险，无论是数据库本身遭受的损坏（机房出现问题、数据库数据被损坏或者篡改），还是中心数据库和外部终端的通信出现问题，都会导致集中式数据库不能发挥作用。而区块链的分布式存储就是把全部记录分布式保存在整个网络的多个记账节点上，单个节点的损坏或丢失并不会对其他节点造成影响，单个节点的数据错误或篡改更不可能对整体数据产生什么破坏性影响。存储在不同节点的数据信息都受加密技术的严格保护，即使获得了相关信息，没有合法授权也无法偷窥到数据信息的真实内容。

区块链通过共识机制创造性地解决了全网各记账节点的信息同步问题，可以有效摆脱某些问题节点的影响，完成正确的记账更新。以比特币为代表的PoW共识机制还能通过提供有效的反馈来激励网络节点参与记账，吸引全世界的算力来充当比特币的账房先生，因而有效维护了整个比特币区块链网络的运行和发展。

Unit Eight Big Data and Artificial Intelligence

8.1 Learning Goals

After learning this unit, you will be able to master the following knowledge:

1. As hardware performance improves and computing resources become more and more powerful, Big Data breeds today's Artificial Intelligence.

2. Unstructured data involves all kinds of information sources, and its proportion will rise sharply in the era of Big Data.

3. Without the help of Artificial Intelligence, Big Data is difficult to do high-level processing and analysis.

4. Artificial Intelligence and Big Data are a model of symbiotic development.

8.2 Dialogue

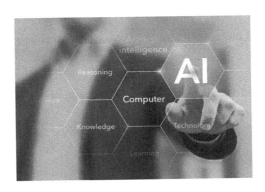

In order to promote the rapid development of the company in the areas of Artificial Intelligence, Baidu in two years ago hired Google Brain team founder Andrew as its chief scientist. Joan, a journalist from CCTV, is interviewing him.

Joan: What makes you start to study Artificial Intelligence?

Andrew: Since the beginning of high school, I have been very interested in Artificial Intelligence. My father is a doctor. When he was young, he has written some Artificial Intelligence algorithms and softwares for medical diagnosis. Therefore, I grew up exposed to the field and began to understand the knowledge of Neural Network.

Joan: It has been attracted to you. Where is it?

Andrew: In fact, we have to spend a lot of time in life completing those boring chores (drive, write a report, and so on). If we can give these work to the machine, we can put the time for things that are interesting. In high school, I think, if the Artificial Intelligence system can help us to do a lot of work, we can spend time staying with our wife and family.

Joan: But, I heard that a lot of people are afraid to give things to robots to complete. For these people, what do you have to say?

Andrew: I have engaged in the study of Artificial Intelligence research in that it is a revolution in technology. Artificial Intelligence will have an impact on many of the posts. However, it will make the whole society better. Many people don't think Artificial Intelligence will bring social great changes. If they are able to advance their thought, society is definitely getting better and better.

Joan: In what areas do you think the depth of learning in the future will mainly be used?

Andrew: At present, I have observed the direction of the application including the unmanned vehicle and speech recognition. Speech recognition may sound unremarkable, but it may bring about changes in society as a whole and thus lead to the social trend. Now, speech recognition technology is not satisfactory. However, I hope we'll have such equipment through the voice of good interaction in the near future.

Joan: What you have said is quite impressive. Thank you very much.

Andrew: Not at all. Good luck, Joan!

8.3 Reading

Artificial Intelligence involves a wide range of fields, and has penetrated into people's lives. Based on the accumulation and application of Big Data, people began to find some rules in it, which triggered the demand for analysis, and made the machine start to have ideas. As hardware performance improves and computing resources become more and more powerful, Big Data breeds today's Artificial Intelligence. Many things that only people can do in the past can be realized by machines gradually. Typical cases include voice assistant, driverless, robot.

AI is based on Big Data, but these resources are usually in the hands of giants. That's why Microsoft, Google, IBM, Apple, Amazon and Facebook are always hitting the headlines in this field. In China, BAT, JD and other enterprises also have enough user base, and have carried out applications.

Artificial Intelligence itself is the partner that Big Data must be combined for further development. Big Data collection is not only standardized data, but also a large number of unstructured data and different dimensions of complex information.

Structured data can be stored in an ordinary database, and the corresponding processing will be more standard and simple. Unstructured data involves all kinds of information sources, and its proportion will rise sharply in the era of Big Data. Especially with the rise of social media, unstructured data ushered in explosive growth. It's not easy to analyse Big Data. Whether it's Natural Language Processing technology, image analysis technology or speech recognition technology, it's a traditional research area of Artificial Intelligence. Without the help of Artificial Intelligence, Big Data is difficult to do high-level processing and analysis. The Deep Learning examples of the Neural Network known as AlphaGo is a further development of Artificial Intelligence. Artificial Intelligence needs not only the analysis and interpretation of existing information, but also the establishment of active information acquisition and learning ability.

Obviously, Artificial Intelligence and Big Data are a model of symbiotic development. With the accumulation of more data and the support of better hardware level, Artificial Intelligence will develop stronger vitality.

8.4 New Words & Phrases & Sentences

8.4.1 New Word

1. symbiotic　　　　　　　　　*adj.* 共生的；互惠互利的
2. artificial　　　　　　　　　*adj.* 人工的；人造的；假的；人为的；非自然的；虚假的；假装的
3. penetrate　　　　　　　　　*v.* 穿过；进入；渗透，打入（组织、团体等）；看透；透过……看见
4. trigger　　　　　　　　　　*v.* 发动；引起；触发；开动；起动
5. dimension　　　　　　　　　*n.* 维（构成空间的因素）；尺寸；规模；程度；范围；方面；侧面
6. breed　　　　　　　　　　　*v.* 交配繁殖；饲养，培育（动植物）；孕育
   n. 品种（尤指人工培育的狗、猫或牲畜）；（人的）类型，种类
7. proportion　　　　　　　　　*n.* 部分；份额；比例；倍数关系；正确的比例；均衡；匀称
8. journalist　　　　　　　　　*n.* 新闻记者；新闻工作者
9. explosive　　　　　　　　　*adj.* 易爆炸的
10. recognition　　　　　　　　*n.* 认出；认识；识别；承认；认可
11. interpretation　　　　　　　*n.* 理解；解释；说明；演绎；演奏方式；表演方式
12. acquisition　　　　　　　　*n.* 获得，得到；购得物；购置物；收购的公司；购置的产业；购置；收购

13.	vitality	*n.* 生命力；活力；热情
14.	unremarkable	*adj.* 一般的；平常的；平凡的；平庸的
15.	usher	*v.* 把……引往；引导；引领

8.4.2 Phrases

1.	high-level processing and analysis	高级处理与分析
2.	promote the rapid development	促进快速发展
3.	in order to	为了
4.	Artificial Intelligence	人工智能
5.	a wide range of	大范围的；许多各种不同的
6.	computing resource	计算资源
7.	typical cases	典型案例
8.	medical diagnosis	医学诊断
9.	Neural Network	神经网络
10.	based on	以……为基础
11.	standardized data	标准化数据
12.	complex information	复杂信息
13.	unstructured data	非结构化数据
14.	engage in	参加；从事；忙于
15.	Natural Language Processing (NLP)	自然语言处理
16.	in the era of Big Data	在大数据时代
17.	impact on	影响；对……冲击，碰撞
18.	lead to the social trend	引领社会潮流
19.	speech recognition technology	语音识别技术
20.	a model of symbiotic development	共生发展模式

8.4.3 Sentences

1. Artificial Intelligence involves a wide range of fields, and has penetrated into people's lives. 人工智能涉及的范围非常大，并且已经渗透到了人们的生活中。

2. Based on the accumulation and application of Big Data, people began to find some rules in it, which triggered the demand for analysis, and made the machine start to have ideas. 在大数据积累和应用的基础上，人们开始在其中寻找某种规律，这引发了对分析的需求，使机器开始有了思想。

3. Many things that only people can do in the past can be realized by machines gradually. Typical cases include voice assistant, driverless, robot. 很多过去只有人能做的事情，现在逐渐能够通过机器实现，典型的案例包括语音助手、无人驾驶、机器人。

4. AI is based on Big Data, but these resources are usually in the hands of giants. AI 的基础是大数据，但这些资源通常掌握在巨头手中。

5. Artificial Intelligence itself is the partner that Big Data must be combined for further development. 人工智能本身就是大数据进一步发展必须结合的伙伴。

6. Structured data can be stored in an ordinary database, and the corresponding processing will be more standard and simple. 结构化数据可以存储在普通数据库中，且相应的处理会比较规范和简单。

7. Especially with the rise of social media, unstructured data ushered in explosive growth. 特别是随着社交媒体的兴起，非结构化数据迎来了爆发式增长。

8. Obviously, Artificial Intelligence and Big Data are a model of symbiotic development. 显然，人工智能和大数据是共生发展模式。

8.5 Exercises

【Ex. 1】 Content Questions.

1. What is the difference of structured data and unstructured data？
2. Why is AI important？

【Ex. 2】补充空白部分的单词。

1. Artificial Intelligence involves a _____ _____ of fields, and has penetrated into people's lives.

2. Based on the _____ and _____ of Big Data, people began to find some rules in it, which _____ the demand for analysis, and made the machine start to have ideas.

3. As hardware performance _____ and computing resources become more and more _____, Big Data breeds today's Artificial Intelligence.

4. AI is based on _____ _____, but these resources are usually in the hands of giants.

5. _____ _____ itself is the partner that Big data must be combined for further development.

6. Big Data collection is not only _____ data, but also a large number of _____ data and different dimensions of complex information.

7. Artificial Intelligence needs not only the _____ and _____ of existing information, but also the establishment of active information acquisition and learning ability.

8. With the _____ of more data and the support of better hardware level, Artificial Intelligence will _____ stronger vitality.

8.6　Translations for Dialogue

为了促进公司在人工智能领域的快速发展，百度在两年前聘请了 Google Brain 团队的创始人安德鲁为首席科学家。中央电视台记者琼正在采访他。

琼：什么使你开始研究人工智能呢？

安德鲁：从高中开始，我就对人工智能非常感兴趣。我父亲是一名医生。年轻时，他曾为医学诊断写过一些人工智能算法和软件。因此，我从小就接触这一领域，并开始了解神经网络的知识。

琼：它吸引你的地方在哪里呢？

安德鲁：事实上，我们在生活中不得不花很多时间来完成那些无聊的杂务（开车、写报告等）。如果我们能把这些工作交给机器，就可以把时间花在有趣的事情上。在高中时我曾想，如果人工智能系统能帮助我们做很多工作，我们就可以花时间和我们的妻子和家人在一起。

琼：但是，我听说很多人都害怕把事情交给机器人来完成。对于这些人，你有什么要说的吗？

安德鲁：我从事人工智能的研究是因为它是一场技术革命。人工智能将对许多职位产生影响。但是，它将使整个社会变得更好。很多人不认为人工智能会给社会带来巨大变化。如果他们能够改进这些思想，社会肯定会越来越好。

琼：你认为未来深度学习将主要用于哪些方面？

安德鲁：目前，我已经观察到了应用的方向，包括无人驾驶汽车和语音识别。语音识别可能听起来平凡，但它可能会给整个社会带来变化，从而引领社会潮流。目前，语音识别技术还不尽如人意。但是，我希望在不久的将来，我们能通过良好的声音交互拥有这样的设备。

琼：你所说令人印象深刻，非常感谢。

安德鲁：不客气。祝你好运，琼！

8.7　Translations for Reading

人工智能涉及的范围非常大，并且已经渗透到了人们的生活中。在大数据积累和应用的基础上，人们开始在其中寻找某种规律，这引发了对分析的需求，使机器开始有了思想。当硬件性能逐渐提高、计算资源越来越强大时，大数据孕育了今天的人工智能。很多过去只有人能做的事情，现在逐渐能够通过机器实现，典型案例包括语音助手、无人驾驶、机器人。

AI 的基础是大数据，但这些资源通常掌握在巨头手中。这也是微软、谷歌、IBM、苹果、亚马逊、Facebook 总是成为该领域的新闻焦点的原因。在国内，BAT（百度、阿里巴巴、腾讯）、京东和其他企业同样拥有足够的用户基础，并且已经开展了应用。

人工智能本身就是大数据进一步发展必须结合的伙伴。大数据采集的不仅是标准化数据，还有大量非结构化数据和不同维度的复杂信息。

结构化数据可以存储在普通数据库中，且相应的处理会比较规范和简单。而非结构化数据涉及各种信息来源，在大数据时代，其比例将急剧上升。特别是随着社交媒体的兴起，非结构化数据迎来了爆发式增长。对大数据进行分析并不简单，无论是自然语言处理技术，还是图像分析技术或语音识别技术，都是传统意义上人工智能的研究领域。没有人工智能的帮助，大数据很难完成高级处理与分析。神经网络的深度学习例子 AlphaGo 是人工智能的进一步发展。人工智能不仅需要分析和解读已有信息，还需要建立主动的信息获取和学习能力。

显然，人工智能和大数据是共生发展模式。随着更多的数据积累和更好的硬件水平支持，人工智能将迸发出更强大的生命力。

Appendix A 49 例大数据术语

1. 大数据（Big Data）

大数据是需要新处理模式才能具有更强的决策力、洞察发现力和流程优化能力的海量、高增长率和多样化的信息资产。

2. 大数据的 5V

Volume（量）、Velocity（速度）、Variety（种类）、Veracity（真实性）、Value（价值）。

3. 生态圈

开源大数据生态圈：

➢ Hadoop HDFS、Hadoop MapReduce、HBase、Hive 依次诞生，早期 Hadoop 生态圈逐渐形成；

➢ Hypertable 不属于 Hadoop 生态圈；

➢ NoSQL，MemBase、MongoDB。

商用大数据生态圈：

➢ IBM PureData (Netezza)、Oracle Exadata、SAP HANA 等；

➢ Teradata Aster Data、EMC Greenplum、HP Vertica 等；

➢ QlikView、Tableau 及国内的 Yonghong Data Mart。

4. Hadoop

Hadoop 是由 Apache 软件基金会开发的分布式系统基础架构。

用户可以在不了解分布式底层细节的情况下开发分布式程序。充分利用 Hadoop 实现高速运算和存储。

Hadoop 实现了分布式文件系统（Hadoop Distributed File System），简称 HDFS。HDFS 具有高容错性，可以部署在价格低廉的硬件中，还可以提供高吞吐量，适合那些有大型数据集（large data set）的应用程序。HDFS 放宽了对 POSIX 的要求，可以以流的形式访问文件系统中的数据。

Hadoop 最核心的设计是 HDFS 和 MapReduce。HDFS 为海量数据提供了存储功能，MapReduce 为海量数据提供了计算功能。

5. Apache 软件基金会

Apache 软件基金会（Apache Software Foundation，ASF），是专门为支持开源软件项目而办的一个非营利性组织。在它所支持的 Apache 项目与子项目中，所发行的软件产品都遵循 Apache 许可证（Apache License）。

6. MapReduce

MapReduce 是一种编程模型，用于大规模数据集（大于 1TB）的并行运算。Map（映射）和 Reduce（归约）的概念和它们的主要思想都来自函数式编程语言，还有矢量编程语言的特性。它极大地方便了编程人员在不会分布式并行编程的情况下，将程序运行在分布式系统中。当前的软件实现是指定一个 Map（映射）函数，使一组键值对映射成一组新的键值对；指定并发的 Reduce（归约）函数，保证所有映射的键值对共享相同的键组。

7. 商业智能（Business Intelligence）

BI（Business Intelligence）即商业智能，它是一套完整的解决方案，可以将企业中现有的数据有效整合，快速准确地提供报表并提出决策依据，帮助企业做出明智的业务经营决策。

8. CRM

CRM 即客户关系管理，企业用 CRM 技术来管理与客户之间的关系。在不同场合中，CRM 可能是一个管理学术语，也可能是一个软件系统。通常 CRM 指用计算机自动化分析销售、市场营销、客户服务及应用等流程的软件系统。它的目标是通过提高客户的价值、满意度、营利性和忠实度来缩减销售周期和销售成本、增加收入、寻找扩展业务所需的新的市场和渠道。CRM 是选择和管理有价值客户及其关系的一种商业策略，CRM 要求通过以客户为中心的企业文化来支持有效的市场营销、销售与服务流程。

9. 云计算（Cloud Computing）

云计算是基于互联网的相关服务的增加、使用和交付模式，通常涉及通过互联网来提供动态易扩展的虚拟化资源。云是网络、互联网的一种比喻说法。过去往往用云来表示电信网，后来也用来表示互联网和底层基础设施的抽象。云计算强大的计算能力可以模拟核爆炸、预测气候变化和市场发展趋势。用户可以通过计算机、手机等接入数据中心，根据自己的需求进行运算。

10. 云计算相关

分布式计算（distributed computing）、并行计算（parallel computing）、效用计算（utility computing）、网络存储（network storage technology）、虚拟化（virtualization）、负载均衡（load balance）、热备份冗余（high available）。

11. 数据仓库

数据仓库是为企业所有级别的决策制定过程提供支持的所有类型数据的战略集合。创建数据仓库的目的是提供分析性报告和决策支持，以指导业务流程改进和监视时间、成本、质量。

12. 非关系型数据库

NoSQL 泛指非关系型数据库。随着互联网 Web2.0 网站的兴起，传统的关系型数据库在应付 Web2.0 网站，特别是超大规模和高并发的 SNS 类型的 Web2.0 纯动态网站时，已经显得力不从心，暴露了很多难以克服的问题，而 NoSQL 数据库则由于其本身的特点而实现了快速发展。NoSQL 数据库的产生就是为了解决大规模数据集和多重数据种类带来的挑战，尤其是大数据应用难题。

13. 结构化数据（structured data）

结构化数据即行数据，存储在数据库里，可以用二维逻辑表来表现。相反，不方便用数据库二维逻辑表来表现的数据即为非结构化数据，包括所有格式的办公文档、文本、图片、标准通用标记语言下的子集 XML、HTML、各类报表、图像、音频和视频信息等。

14. 结构化分析方法（structured method）

结构化分析方法是强调开发方法的结构合理性及所开发软件的结构合理性的软件开发方法。结构是指系统内各组成要素之间相互联系、相互作用的框架。结构化开发方法提出了一组提高软件结构合理性的准则，如分解与抽象、模块独立性、信息隐蔽等。针对软件生存周期的不同阶段，有结构化分析（SA）和结构化程序设计（SP）等方法。

15. 半结构化数据（semi-structured data）

和普通纯文本相比，半结构化数据具有一定的结构性。OEM（Object Exchange Model）是一种典型的半结构化数据模型。

在进行信息系统设计时肯定会涉及数据的存储，一般将系统信息保存在某个指定的关系型数据库中。按业务对数据进行分类，并设计相应的表，然后将对应的信息保存到相应的表中。例如，我们做一个业务系统，保存员工的基本信息：工号、姓名、性别、出生日期等，就可以建立一个对应的 staff 表。但不是系统中的所有信息都可以简单地用一个表中的字段来对应。

16. 非结构化数据库

非结构化数据库是指字段长度可变，并且每个字段的记录都可以由可重复或不可重复的子字段构成的数据库，不仅可以用来处理结构化数据（数字、符号等信息），还可以用来处理非结构化数据（全文文本、图像、声音、影视、超媒体等信息）。

17. 数据库（database）

数据库是按照数据结构来组织、存储和管理数据的仓库，它产生于 60 多年前。随着信息技术和市场的发展，特别是 20 世纪 90 年代以后，数据管理不再只是存储和管理数据，而是转变成了用户所需要的各种数据管理方式。数据库有多种类型，从最简单的存储各种数据的表格，到能够存储海量数据的大型数据库系统，应用广泛。

18. 数据分析（data analysis）

数据分析是指用适当的统计分析方法对收集来的大量数据进行分析，是为了提取有用信息和形成结论而对数据进行详细研究和概括总结的过程。这一过程也是质量管理体系的支持过程。在实用中，数据分析可以帮助人们做出判断，以采取适当行动。

19. 数据挖掘（data mining）

数据挖掘是知识发现（Knowledge Discovery in Databases，KDD）中的步骤。数据挖掘一般是指从大量数据中通过算法搜索隐藏信息的过程。数据挖掘通常与计算机科学有关，通过统计、在线分析处理、情报检索、机器学习、专家系统（依靠过去的经验法则）和模式识别等方法来实现上述目标。

20. 数据清洗（data cleansing）

数据清洗指发现并纠正数据文件中可识别的错误的最后一道程序，包括检查数据一致性及处理无效值和缺失值等。由于数据仓库中的数据是面向某一主题的数据的集合，这些数据从多个业务系统中抽取而来且包含历史数据，所以无法避免存在错误数据和冲突数据，这些错误的或有冲突的数据显然是我们不想要的，称其为"脏数据"。我们要按照一定的规则把"脏数据"洗掉，即数据清洗。数据清洗的任务是过滤那些不符合要求的数据，将过滤的结果交给业务主管部门，确认是否过滤掉并由业务单位修正后再抽取。不符合要求的数据主要包括不完整的数据、错误的数据、重复的数据。数据清洗与问卷审核不同，录入后的数据清洗一般由计算机完成，而不是由人工完成。

21. 可视化（visualization）

可视化是利用计算机图形学和图像处理技术，将数据转换成图形或图像在屏幕上显示出来，并进行交互处理的理论、方法和技术。它涉及计算机图形学、图像处理、计算机视觉、计算机辅助设计等领域，是研究数据表示、数据处理、决策分析等一系列问题的综合技术。目前正在飞速发展的虚拟现实技术也是以图形图像的可视化技术为依托的。

22. 数据可视化（Data Visualization）

数据可视化技术的基本思想是将数据库中的每个数据项作为单个图元元素表示，大量数据集构成数据图像，同时将数据的各属性值以多维数据的形式表示，可以从不同的维度观察数据，从而对数据进行更深入的观察和分析。

数据可视化旨在借助图形化手段，清晰有效地传达信息。但是，这并不意味数据可视化因为要实现其功能而令人感到枯燥乏味，或者为了看上去绚丽多彩而显得极端复杂。为了有效地传达思想观念，美学形式与功能需要齐头并进，通过直观地传达关键方面与特征，实现对稀疏而复杂的数据集的深入洞察。然而，设计人员往往不能很好地实现设计与功能的平衡，导致其创造出华而不实的数据可视化形式，无法达到其主要目的，即传达和沟通信息。

23. 产品数据管理（Product Data Management，PDM）

产品数据管理是基于分布式网络、主从结构、图形化用户接口和数据库管理技术发展起来的一种软件框架（或数据平台），PDM 对并行工程中的人员、工具、设备、资源、产品数据及数据生成过程进行全面管理。

24. 需求方平台（Demand Side Platform，DSP）

需求方平台的概念起源于网络广告发达的欧美国家，是随着互联网和广告业的飞速发展而新兴起的网络广告领域。它与 Ad Exchange 和 RTB 一同在美国崛起，并在全球快速发展。DSP 传入中国后迅速引发热潮，推动中国网络展示广告市场快速发展。

25. 数据管理平台（Data Management Platform，DMP）

数据管理平台将分散的第一方、第二方和第三方数据整合并纳入统一的技术平台，并对这些数据进行标准化和细分，让用户可以将这些细分结果推入现有的互动营销环境中。

DMP 的核心元素如下。

- ➢ 数据整合及标准化能力：采用统一的方式，吸纳和整合各方数据；
- ➢ 数据细分管理能力：创建独一无二、有意义的客户细分，进行有效营销；

➢ 功能健全的数据标签：提供灵活的数据标签，便于开展营销活动；

➢ 自助式用户界面：基于 Web 界面或其他集成方案直接获取数据工具、功能和几种形式的报表；

➢ 相关渠道环境的连接：与相关渠道集成，包含网站端、展示广告、电子邮件及搜索和视频，让营销者能找到、定位和提供与细分群体高度相关的营销信息。

26. CPA（Cost Per Action）

CPA 是一种广告计费模式，将行为作为计费指标，该行为可以是注册、咨询、放入购物车等。广告公司和媒体公司常用 CPA、CPC（Cost Per Click）、CPM（Cost Per Mille）来衡量广告价格。

CPA 计价方式是指按广告投放实际效果，即按回应的有效问卷或订单来计费，而不限广告投放量。CPA 广告是网络中最常见的广告形式之一，当用户点击某网站上的广告后，这个站的站长就会获得相应的收入。

27. CPT（Cost Per Time）

按时长计费是一种包时段投放广告的形式，广告主选择广告位和投放时间，费用与广告点击量无关。网站主决定每个广告位的价格，广告主自行选择购买时间段，目前可按周或按天购买，成交价就是网站主标定的价格。

28. CTR（Click Through Rate）

CTR 是互联网广告常用的术语，指网络广告（图片广告、文字广告、关键词广告、排名广告、视频广告等）的点击到达率，即该广告的点击量（严格来讲，可以是到达目标页面的数量）与广告浏览量（Page View，PV）的比。

CTR 是衡量互联网广告效果的重要指标之一。

29. 算法（algorithm）

算法是对解题方案准确而完整的描述，是一系列用于解决问题的清晰指令，算法代表用系统的方法描述解决问题的策略机制。也就是说，能够在有限时间内针对具有一定规范的输入获得所要求的输出。如果一个算法有缺陷，或不适用于某个问题，那么执行这个算法将不会解决这个问题。不同的算法可能用不同的时间、空间或效率完成相同的任务。算法的优劣可以用空间复杂度与时间复杂度来衡量。

30. 机器学习（Machine Learning，ML）

机器学习是一门多领域交叉学科，涉及概率论、统计学、逼近论、凸分析、算法复杂度理论等多门学科。专门研究计算机怎样模拟或实现人类的学习行为，以获取新的知识或技能，并重新组织已有的知识结构，不断改善自身性能。

机器学习是人工智能的核心，是使计算机具有智能的根本途径，其应用遍及人工智能各领域，它主要使用归纳、综合而不是演绎。

31. 人工智能（Artificial Intelligence，AI）

人工智能是研究和开发用于模拟、延伸和扩展人类智能的理论、方法、技术及应用系统的一门新的技术科学。人工智能是计算机科学的一个分支，它企图了解智能的实质，并生产一种能以与人类智能相似的方式做出反应的智能机器，该领域的研究有机器人、语言识别、图像识别、自然语言处理和专家系统等。自诞生以来，人工智能的理论和技术日益成熟，应用领域不断扩大。可以设想，未来人工智能带来的科技产品，将会是人类智慧的"容器"。

32. 深度学习（Deep Learning，DL）

深度学习的概念源自对人工神经网络的研究。多层感知器就是一种深度学习结构。深度学习通过组合低层特征形成更加抽象的高层特征，以发现数据的分布式特征。

深度学习的概念由 Hinton 等人于 2006 年提出。他们基于深度置信网络（DBN）提出了非监督贪心逐层训练算法，为解决与深层结构相关的优化难题带来希望，随后提出多层自动编码器深层结构。此外，LeCun 等人提出的卷积神经网络是第一个真正的多层结构学习算法，它利用空间相对关系减少了参数数量，提高了训练性能。

深度学习是机器学习研究的新领域，其动机在于建立、模拟人脑进行分析学习的神经网络。它通过模仿人脑的机制来解释图像、声音和文本数据。

33. 神经网络

人工神经网络（Artificial Neural Network，ANN）简称神经网络或连接模型（Connection Model），它是一种模仿动物神经网络行为特征进行分布式并行信息处理的算法数学模型。这种网络根据系统的复杂程度，通过调整内部大量节点之间相互连接的关系，来达到处理信息的目的。

34. OpenStack

OpenStack 由 NASA（美国国家航空航天局）和 Rackspace 合作研发，是经 Apache 许可证授权的自由软件和开源项目。

OpenStack 是一个开源的云计算管理平台项目，由几个主要的组件组合完成具体工作。OpenStack 几乎支持所有类型的云环境，项目目标是提供实施简单、可大规模扩展、丰富、标准统一的云计算管理平台。OpenStack 通过各种互补的服务提供了基础设施即服务（IaaS）解决方案，并为每项服务提供 API 以进行集成。

35. SaaS

SaaS 是软件即服务（Software as a Service）。随着互联网技术的发展和应用软件的成熟，21 世纪开始兴起一种完全创新的软件应用模式，即 SaaS。与按需软件（on-demand software）、应用服务供应商（application service provider）、托管软件（hosted software）的含义相似。它是一种通过 Internet 提供软件的模式，厂商将应用软件统一部署在自己的服务器上，客户可以根据自己实际需求，通过互联网向厂商订购所需的应用软件服务，按订购的服务数量和时间长短向厂商支付费用，并通过互联网获得厂商提供的服务。

36. PaaS

PaaS 是平台即服务（Platform as a Service）。它将服务器平台作为一种提供服务的商业模式。通过网络提供程序的服务被称为 SaaS，而在云计算时代，将相应的服务器平台或开发环境作为服务进行提供则为 PaaS。

PaaS 指将软件研发平台（计世资讯将其定义为业务基础平台）作为一种服务，以 SaaS 的模式提交给用户。因此，PaaS 也是 SaaS 模式的一项应用。但是，PaaS 的出现可以加快 SaaS 的发展，尤其是加快 SaaS 应用的开发速度。2007 年，国内外 SaaS 厂商先后推出了自己的 PaaS 平台。

37. IaaS

IaaS 是基础设施即服务（Infrastructure as a Service）。消费者可以通过互联网从完善的计算机基础设施获得服务，这类服务被称为 IaaS。基于互联网的服务（如存储和数据库）是 IaaS 的一部分。互联网上其他类型的服务包括平台即服务（Platform as a Service，PaaS）和软件即服务（Software as a Service，SaaS）。PaaS 提供了用户可以访问的完整或部分应用程序开发，SaaS 则提供了可直接使用的完整应用程序，如通过互联网管理企业资源。

38. HaaS

HaaS 是硬件即服务（Hardware as a Service）。HaaS 概念源于云计算，现在被称为基础架构即服务（IaaS）或基础架构云，各企业可以使用 IaaS 通过互联网将更多的基础架构容量作为服务提供。通过互联网分配更多的存储或处理容量比供应商在基础环境中引入和安装新硬件要快得多。HaaS 还具有一层针对嵌入式设备的含义，目的在于建立通过互联网进行嵌入式设备统一管理服务的模式。在这种情况下，HaaS 与 SaaS 类似，对于嵌入式设备使用者来说，无须一次性购买所需嵌入式设备，仅需按照设备使用量或其他标准支付设备的服务费及维护费即可。

39. 决策树（Decision Tree）

决策树是在已知各种情况发生概率的基础上，通过构成决策树来求取净现值的期望值大于等于零的概率、评价项目风险、判断其可行性的决策分析方法，是一种直观运用概率分析的图解法。由于这种决策分支画成图形很像一棵树的枝干，故称其为决策树。在机器学习中，决策树是一个预测模型，代表对象属性与对象值之间的一种映射关系。

40. EM 算法

最大期望算法即 EM 算法（Expectation Maximization Algorithm），是一种迭代算法，用于含有隐变量（hidden variable）的概率参数模型的最大似然估计或极大后验概率估计。

41. 聚类分析（cluster analysis）

聚类分析是静态数据分析技术，在许多领域得到了广泛应用，如机器学习、数据挖掘、模式识别等。聚类分析通过静态分类方法将相似的对象分成不同的组别或子集（subset），在同一子集中的成员对象存在一些相似性。

42. 概率模型

给定一个用户的查询串，相对于该串存在一个包含所有相关文档的集合。我们把这样的集合看作理想的结果文档集，在给出理想的结果文档集后，我们很容易能得到结果文档。这样就可以把查询处理看作对理想的结果文档集属性的处理。问题是我们并不能准确地了解这些属性，我们了解的是表示这些属性的索引术语。由于在查询期间这些属性都是不可见的，所以需要在初始阶段估计这些属性。这种估计允许我们对首次检索的文档集合返回理想的结果文档集，并产生初步的概率描述。

43. 贝索斯定律（Bezos' Law）

贝索斯定律是指在云的发展过程中，单位计算能力的价格大约每 3 年会降低 50%。

44. 回归分析（regression analysis）

回归分析是确定两种或两种以上变量间相互依赖的定量关系的一种统计分析方法。其运用十分广泛。按照涉及自变量的多少，可以将回归分析分为一元回归分析和多元回归分析；按照自变量和因变量之间的关系，可以将回归分析分为线性回归分析和非线性回归分析。如果在回归分析中，只包括一个自变量和一个因变量，且二者的关系可用一条直线近似表示，则可以称这种回归分析为一元线性回归分析。如果回归分析中包括两个或两个以上自变量，且因变量和自变量之间是线性关系，则称其为多元线性回归分析。

45. 推荐算法

基于内容的信息推荐方法的理论依据主要来自信息检索和信息过滤。基于内容的推荐方法就是根据用户的浏览记录向其推荐没有接触过的推荐项。主要用两个方法来描述基于内容的推荐方法：启发式方法和基于模型的方法。启发式方法指用户凭借经验来定义相关的计算公式，根据公式的计算结果对实际结果进行验证，不断修改公式以达到最终目的；基于模型的方法将以往的数据作为数据集，根据这个数据集来学习模型。

46. 八叉树（octree）

八叉树是一种用于描述三维空间的树状数据结构。八叉树的每个节点表示一个正方体的体积元素，每个节点有 8 个子节点，将 8 个子节点所表示的体积元素加在一起就得到了父节点的体积。

47. 红黑树（Red Black Tree）

红黑树是一种自平衡二叉查找树，是在计算机科学中用到的一种数据结构，典型的用途是实现关联数组。

1972 年，Rudolf Bayer 发明了平衡二叉 B 树（symmetric binary B-tree）。1978 年，Leo J. Guibas 和 Robert Sedgewick 将其修改为红黑树。

红黑树与 AVL 树类似，都是在进行插入和删除操作时通过特定操作保持二叉查找树的平衡，以获得较高的查找性能。

红黑树虽然很复杂，但它的最坏情况运行时间非常好，并且在实践中十分高效：它可以在 $O(\log n)$ 时间内进行查找、插入和删除，这里的 n 是树中元素的数量，$O(\log n)$ 是时间复杂度。

48. 哈希表（Hash Table）

哈希表，也称散列表，是根据键值直接进行访问的数据结构。也就是说，它通过把键值映射到表中的一个位置来访问记录，以加快查找的速度。这个映射函数为散列函数，存放记录的数组为散列表。

给定表 M，存在函数 $f(key)$，若将任意给定的键值 key 代入函数后能得到包含该关键字的记录在表中的地址，则称表 M 为哈希表，称 $f(key)$ 为哈希函数。

49. 随机森林（Random Forest）

在机器学习中，随机森林是包含多个决策树的分类器，由 Leo Breiman 和 Adele Cutler 提出，"Random Forests" 由 1995 年贝尔实验室的 Tin Kam Ho 提出的随机决策森林（Random Decision Forests）发展而来，是 Leo Breiman 的"Bootstrap Aggregating"和 Tin Kam Ho 的"Random Subspace Method"的结合。

Appendix B 总词汇表

A

aggregation

聚合（表示搜索、合并、显示数据的过程。）

algorithm

算法（指可以完成某种数据分析的数学公式。）

analytics

分析法（用于发现数据的内在含义。）

anomaly detection

异常检测（在数据集中搜索与预期模式或行为不匹配的数据项。除了"Anomalies"，用来表示异常的词还有 outliers、exceptions、surprises、contaminants。他们通常用来提供关键的可执行信息。）

anonymization

匿名化（使数据匿名，即移除所有与个人隐私相关的数据。）

application

应用（实现某种特定功能的计算机软件。）

Artificial Intelligence（AI）

人工智能（研发智能机器和智能软件，这些智能设备能够感知周遭环境，并根据要求做出反应，甚至能自我学习。）

B

behavioural analytics

行为分析（行为分析根据用户的行为如"怎么做""为什么这么做"及"做了什么"得出结论，其不仅是一门针对人物和时间的分析学科，还着眼于数据中的人性化模式。）

Big Data scientist

大数据科学家（能够设计大数据算法以使大数据变得有用的人。）

Big Data startup

大数据创业公司（指研发最新大数据技术的新兴公司。）

biometrics

生物计量学（根据个人的特征进行身份识别。）

BB: BrontoByte

千亿亿亿字节

Business Intelligence（BI）

商业智能（是一系列理论、方法学和过程，使得数据更容易被理解。）

C

classification analysis

分类分析（从数据中获得重要相关性信息的系统化过程；这类数据也被称为元数据，是描述数据的数据。）

Cloud Computing

云计算（构建在网络上的分布式计算系统，数据存储在云端。）

cluster analysis

聚类分析（指将相似的对象聚合在一起进行分析。）

cold data storage

冷数据存储（在低功耗服务器上存储那些几乎不被使用的旧数据。但这些数据检索起来会很耗时。）

comparative analysis

对比分析（在非常大的数据集中进行模式匹配时，一步步进行对比和计算，得到分析结果。）

complex structured data

复杂结构化数据（由两个或多个复杂而相互关联的部分组成的数据，这类数据不能简单地由结构化查询语言或工具 SQL 解析。）

computer generated data

计算机生成的数据（如日志文件等由计算机生成的数据。）

concurrency

并发（同时执行多个任务或运行多个进程。）

correlation analysis

相关分析（一种数据分析方法，用于分析变量之间是否存在正相关或负相关关系。）

CRM

客户关系管理（Customer Relationship Management，一种用于管理销售、业务过程的技术。大数据影响公司的客户关系管理策略。）

D

dashboard

仪表板（使用算法分析数据，并以图表方式将结果显示在仪表板中。）

data aggregation tool

数据聚合工具（将分散于众多数据源的数据转化成一个全新数据源的过程。）

Data Analyst

数据分析师（从事数据分析、建模、清洗、处理的专业人员。）

database

数据库（一个以某种特定技术储存数据集合的仓库。）

DaaS

数据库即服务（Database as a Service，部署在云端的数据库，即用即付，如亚马逊云服务 AWS。）

DBMS

数据库管理系统（Database Management System，收集、存储数据，并提供数据访问。）

data centre

数据中心（一个实体地点，放置用来存储数据的服务器。）

data cleansing

数据清洗（对数据进行重新审查和校验的过程，目的在于删除重复信息、纠正存在的错误，并使数据具有一致性。）

Data Custodian

数据管理员（负责维护数据存储所需技术环境的专业技术人员。）

data ethical guideline

数据道德准则（数据道德准则有助于组织机构使其数据透明化，保证数据的简洁、安全。）

data feed

数据订阅（一种数据流，如 Twitter 订阅和 RSS。）

data marketplace

数据集市（进行数据集买卖的在线交易场所。）

data mining

数据挖掘（从数据集中发掘特定模式或信息的过程。）

data modelling

数据建模（使用数据建模技术分析数据对象，以洞悉数据的内在。）

data set

数据集（大量数据的集合。）

Data Virtualization

数据虚拟化（数据整合的过程，可以获得更多数据信息，这个过程通常会引入其他技术，如数据库、应用程序、文件系统、网页技术、大数据技术等。）

de-identification

去身份识别（也称为匿名，确保个人不会通过数据被识别。）

discriminant analysis

判别分析（将数据分类，按不同的分类方式，可将数据分配到不同的群组、类别或目录中。判别分析是一种统计分析法，可以对数据中某些群组或集群的已知信息进行分析，并从中获取分类规则。）

Distributed File System

分布式文件系统（提供简化的高可用方式来存储、分析、处理数据。）

document database

文档数据库（是为存储、管理、恢复文档数据而专门设计的数据库。）

E

Exploratory Data Analysis（EDA）

探索性数据分析（在没有标准的流程或方法的情况下从数据中发掘模式，能够发掘数据和数据集的主要特性。）

EB: ExaByte

艾字节

ETL

数据仓库技术（Extract Transform Load，是将业务系统的数据抽取、清洗、转换后，加载到数据仓库的过程。能够将企业中的分散、零乱、标准不统一的数据整合到一起，为企业的决策提供分析依据。）

F

failover

故障切换（当系统中的一个服务器发生故障时，自动将运行任务切换到另一个可用的服务器或节点上。）

fault-tolerant design

容错设计(一个支持容错设计的系统应该能够做到当某部分出现故障时也能继续运行。)

G

gamification

游戏化(指在非游戏领域中运用游戏的思维和机制,这种方法可以以一种十分友好的方式创建和侦测数据,非常有效。)

graph database

图数据库(使用图结构来存储数据,如一组有限的有序对或某种实体,这种图结构包括边缘、属性和节点。它提供了相邻节点间的自由索引功能,也就是说,数据库中的每个元素都与其他相邻元素直接关联。)

grid computing

网格计算(将许多分布在不同地点的计算机连接在一起,用以处理某个特定问题,通常通过云将计算机相连在一起。)

H

Hadoop

一个开源的分布式系统基础框架(可用于开发分布式程序,进行大数据的运算与存储。)

HBase

Hadoop 的数据库(一个开源非关系型分布式数据库,与 Hadoop 框架一同使用。)

HDFS

Hadoop 分布式文件系统(Hadoop Distributed File System,是一个被设计成适合在通用硬件上运行的分布式文件系统。)

HPC

高性能计算(High Performance Computing,使用超级计算机来解决复杂的计算问题。)

I

IMDBMS

内存数据库管理系统(In-Memory Database Management System,一种数据库管理系统,

与普通数据库管理系统的区别在于，它用主存来存储数据，而非硬盘。其特点在于能高速处理和存取数据。）

Internet of Things

物联网

J

juridical data compliance

法律上的数据一致性（当你使用的云计算解决方案将你的数据存储在不同国家时，就会与这个概念扯上关系了。你需要留意这些存储在不同国家的数据是否符合当地法律规定。）

K

key-value database

键值数据库（数据的存储方式是使用一个特定的键，指向一个特定的数据记录，这种方式使数据的查找更加方便快捷。键值数据库中存储的数据通常为编程语言中具有基本数据类型的数据。）

L

latency

延迟（表示系统时间的延迟。）

legacy system

遗留系统（旧的应用程序、旧的技术或旧的计算系统。）

load balance

负载均衡（将工作量分配到多台计算机或服务器上，以获得最优的结果和最大的系统利用率。）

location data

位置数据

log file

日志文件（由计算机系统自动生成的文件，记录系统的运行过程。）

M

Machine to Machine（M2M）

数据算法模型（指数据从一台终端传送到另一台终端，即机器与机器的对话。）

machine data

机器数据（由传感器或算法在机器上产生的数据。）

Machine Learning（ML）

机器学习（人工智能的一部分，指机器能够从它们完成的任务中自我学习，通过长期积累实现自我改进。）

MapReduce

一种处理大规模数据的软件框架

MPP

大规模并行处理（Massively Parallel Processing，同时使用多个处理器或多台计算机处理同一个计算任务。）

metadata

元数据（描述数据的数据，是描述数据属性的信息。）

MongoDB

一种开源的非关系型数据库（旨在为网络应用提供可扩展的高性能数据存储解决方案。）

Multi-Dimensional Database

多维数据库（用于优化数据联机分析处理程序和数据仓库的数据库。）

MultiValue Database

多值数据库（一种非关系型数据库，也是一种特殊的多维数据库，能处理 3 个维度的数据，能够完美处理 HTML 和 XML 中的字符串。）

N

Natural Language Processing（NLP）

自然语言处理（是计算机科学的分支领域，研究如何实现计算机与人类语言的交互。）

network analysis
网络分析（分析网络或图论中节点间的关系，即分析网络中节点间的连接和强度关系。）

NewSQL
一类新式的关系型数据库管理系统（比 SQL 更易学习和使用，比 NoSQL 提出的更晚。）

NoSQL
泛指非关系型数据库（这类数据库有更强的一致性，能处理超大规模和高并发的数据。）

O

Object-Oriented Database
面向对象数据库（以对象的形式存储数据，用于面向对象的编程。它与关系型数据库和图数据库不同，大部分面向对象数据库都提供一种查询语言，允许使用声明式编程访问对象。）

Object-Based Image Analysis
基于对象的图像分析（数字图像分析对所有像素的数据进行分析，而基于对象的图像分析只分析相关像素的数据，这些相关像素被称为对象或图像对象。）

Operational Database
操作型数据库（这类数据库可以完成一个组织机构的常规操作，对商业运营来说非常重要，一般使用在线事务处理，允许用户访问、收集、检索企业内部的具体信息。）

optimization analysis
优化分析（在进行产品设计时，依靠算法来实现优化过程，在这一过程中，公司可以设计各种各样的产品并测试这些产品是否满足预设值。）

ontology
本体论（表示知识本体，用于定义一个领域中的概念集及概念之间的关系。）

outlier detection
异常值检测（异常值是指严重偏离一个数据集或一个数据组合总平均值的对象，该对象与数据集中的其他对象相去甚远。因此，异常值的出现意味着系统发生问题，需要对此另做分析。）

P

Pattern Recognition

模式识别(通过算法识别数据中的模式,并对同一数据源中的新数据进行预测。)

PB: PetaByte

拍字节

PaaS

平台即服务(Platform as a Service,一种为云计算解决方案提供所有必需的基础平台的服务。)

predictive analysis

预测分析(大数据分析方法中最有价值的分析方法之一,这种方法有助于预测个人的未来行为,如某人可能会购买某些商品、访问某些网站、做某些事情或产生某种行为。通过使用各种不同的数据集来识别风险和机遇,如历史数据、事务数据、社交数据和客户的个人信息数据。)

privacy

隐私(把可识别出个人信息的数据与其他数据分离开,以保护用户隐私。)

public data

公共数据(由公共基金创建的公共信息或公共数据集。)

Q

Quantified Self

量化自我(使用应用程序跟踪用户一天的行动,从而更好地理解其相关行为。)

query

查询(查找与某个问题的答案有关的信息。)

R

Re-identification

重新识别(将多个数据集合并在一起,从匿名数据中识别出个人信息。)

regression analysis

回归分析（确定两个变量间的依赖关系。假设两个变量之间存在单向的因果关系。）

RFID

射频识别（使用一种无线非接触式射频电磁场传感器来传输数据。）

real-time data

实时数据（几毫秒内被创建、处理、存储、分析并显示的数据。）

recommendation engine

推荐引擎（根据用户的购买行为向用户推荐某种产品。）

Path analysis

路径分析（常用的数据挖掘方法之一，通过对 Web 服务器的日志文件中客户访问站点访问次数的分析，挖掘频繁访问路径。）

S

semi-structured data

半结构化数据（半结构化数据并不具有结构化数据的严格存储结构，但它可以使用标签或其他形式的标记方式，以保证数据的层次结构。）

sentiment analysis

情感分析（通过算法分析人们是如何看待某些话题的。）

signal analysis

信号分析（指通过度量随时间或空间变化的物理量来分析产品的性能。）

similarity search

相似性搜索（在数据库中查询最相似的对象，这里所说的对象可以是任意类型的数据。）

simulation analysis

仿真分析（仿真是指模拟真实环境中的进程或系统的操作。仿真分析可以在仿真时考虑多种变量，确保产品性能达到最优。）

Smart Grid

智能电网

SaaS

软件即服务（Software as a Service，一种通过浏览器使用的基于互联网的应用软件。）

spatial analysis

空间分析（分析地理信息或拓扑信息等空间数据，从中得出分布在地理空间中的数据模式和规律。）

SQL

结构化查询语言（在关系型数据库中，用于检索数据的一种编程语言。）

structured data

结构化数据（可以组织成行列结构的可识别的数据。这类数据通常是记录、文件，或被正确标记过的数据中的某个字段，可以被精确定位。）

T

TB: TeraByte

太字节

time series analysis

时序分析（分析在重复测量时间内获得的定义好的数据。分析的数据必须是定义好的，并且要取自相同时间间隔的连续时间点。）

Topological Data Analysis（TDA）

拓扑数据分析（拓扑数据分析主要关注 3 点：复合数据模型、集群的识别及数据的统计学意义。）

transactional data

交易数据（随时间变化的动态数据。）

transparency

透明性，透明度（消费者想要知道他们的数据有什么作用、被如何处理，但组织机构则把这些信息都透明化了。）

U

unstructured data

非结构化数据（非结构化数据一般是大量纯文本数据，其中可能包含日期、数字和实例。）

V

value

价值（所有可用的数据能为组织机构、社会、消费者创造巨大价值。这意味着各大企业及整个产业都将从大数据中获益。）

variability

可变性（数据的含义总是在快速变化。例如，一个词在相同的推文中可以有完全不同的意思。）

variety

种类（数据总是以不同的形式呈现，如结构化数据、半结构化数据、非结构化数据和复杂结构化数据。）

velocity

速度（在大数据时代，要快速完成数据的创建、存储、分析、虚拟化过程。）

veracity

真实性（组织机构需要确保数据的真实性，才能保证数据分析的正确性。）

visualization

可视化（可视化不仅指普通的图形或饼图，还包括复杂的图表，图表中包含大量可以被理解和阅读的数据信息。）

volume

量（数据量大。）

W

weather data

天气数据（一种重要的开放公共数据来源，与其他数据来源结合，可以为相关组织机构提供深入分析的依据。）

X

XML database

XML 数据库（XML 数据库是一种以 XML 格式存储数据的数据库。XML 数据库通常与面向文档的数据库关联，开发人员可以对 XML 数据库的数据进行查询、导出及序列化。）

Y

YB: YottaByte

尧字节

Z

ZB: ZettaByte

泽字节

Appendix C 书后练习答案

Unit One

【Ex. 1】Content Questions.

1. According to McKinsey, Big Data refers to data sets whose size is beyond the ability of typical database software tools to capture, store, manage and analyse. There is no explicit definition of how big a data set should be. New technology needs to be in place to manage this Big Data phenomenon.

2. Volume, velocity, variety, veracity, value.

3. The technologies to combine and interrogate Big Data have matured to a point where their deployments are practical. The underlying cost of the infrastructure to power the analysis has fallen dramatically, making it economical to mine the information. The competitive pressure on businesses has increased to the point where most traditional strategies are offering only marginal benefits. Big Data has the potential to provide new forms of competitive advantage for businesses.

4. Value is defined as the usefulness of data for an enterprise. The value characteristic is intuitively related to the veracity characteristic in that the higher the data fidelity, the more value it holds for the business. Value is also dependent on how long data processing takes, because analytics results have a shelf-life. For example, a 20 minutes delayed stock quote has little to no value for making a trade compared to a quote that is 20 milliseconds old.

【Ex. 2】句子翻译。

1. 我希望这次谈话能让你对我们的工作有所了解。

2. 新系统已针对 Microsoft Windows 运行进行了优化。

3. 这些设计展示了她对色彩和细节的敏锐眼光。

4. 让我明确一点：条形图不是分析学。

5. 一本好词典能为我们解释一个词的内涵和外延。

6. 生活方式的最新趋势是慢生活。

7. 一个时代的结束预示着另一个时代的开始。

8. 不能在同一函数中组合结构化和非结构化异常处理。

9. 最后，通过实际应用验证了该方法的可行性和准确性。

10. 多层交换机的可行性取决于其所支持的协议。

【Ex. 3】短文翻译。

1. 云计算是涉及通过互联网交付的托管服务的通用术语。这些服务大致分为3类：基础设施即服务（SaaS）、平台即服务（PaaS）和软件即服务（SaaS）。云计算这个名字的灵感来自经常在流程图和图表中用来表示互联网的云符号。

2. 云服务有3个与传统服务相区别的特征。它是按需出售的，通常按分钟或小时出售；它具有弹性——用户可以在任何给定的时间获得尽可能多或尽可能少的服务；且服务完全由供应商管理（消费者只需要一台个人计算机并接入互联网）。虚拟化和分布式计算的重大创新，以及高速互联网接入的改善和疲软的经济，增加了人们对云计算的兴趣。

Unit Two

【Ex. 1】Content Questions.

1. （1）Continuous growth of digital content. 数字内容的持续增长。

   （2）Proliferation of the Internet of Things (IoT). 物联网的扩散。

   （3）Strong open source initiatives. 强大的开源计划。

   （4）Increasing investments in Big Data technologies. 加大对大数据技术的投资。

   （5）Data Visualization driven by the information-based economy. 信息化经济驱动下的数据可视化。

2. The convergence of mobile devices, the Mobile Internet and Social Networks provide an opportunity for businesses to derive competitive advantage through an efficient analysis of unstructured data. Businesses that were early adopters of Big Data technologies and that based their business on data-driven decision making were able to achieve greater productivity of up to 5% or 6% higher than the norm. 移动设备、移动互联网和社交网络的融合为组织提供了一个通过有效分析非结构化数据来获得竞争优势的机会。早期采用大数据技术的企业和基于数据驱动制定业务决策的企业能够实现比常规企业高5%或6%的生产率。

3. Many of the technologies within the Big Data ecosystem have an open source origin, due to

participation, innovation and sharing by commercial providers in open source development projects. The Hadoop framework, in conjunction with additional software components such as the open source R language and a range of open source database (Not Only Structured Query Language database such as Cassandra and Apache HBase). 由于商业供应商参与、创新和共享开源开发项目，大数据生态系统中的许多技术都是开源的。Hadoop 框架与其他软件组件（如开源 R 语言）和一系列开源数据库（NoSQL 数据库，如 Cassandra 和 Apache HBase）结合。

4. Previously, market information was largely made available through traditional market research and data specialists. Today, virtually any company with a large data sets can potentially become a serious player in the new information game. The value of Big Data will become more apparent to corporate leadership as companies seek to become more "data-driven" organizations. 过去，市场信息主要通过传统的市场研究和数据专家获得。如今，几乎任何拥有大型数据集的企业都有可能成为新信息游戏的重要参与者。随着企业力图成为更大的"数据驱动"型企业，对于公司领导层来说，大数据的价值将变得更加明显。

【Ex. 2】判断以下叙述的正误。

1. 对　　2. 对　　3. 错　　4. 对
5. 错　　6. 错　　7. 对　　8. 错

【Ex.3】选择填空。

（1）C　　（2）A　　（3）A　　（4）D　　（5）C

【Ex. 4】根据给出的汉语词义和规定的词类写出相应的英语单词，每词的首字母已给出。

1. formula	2. redundant	3. deployment
4. scatter	5. batch	6. serve
7. transform	8. inconsistent	9. pattern
10. online	11. repository	12. disparate
13. inaccurate	14. fundamental	15. multidimensional
16. exponent	17. robust	18. functionality
19. scalable	20. implement	

【Ex.5】句子翻译。

1. 你可以和其他人在线聊天。

2. 每年夏天都有新一批学生试图找工作。

3. 你们的产品和服务有竞争力吗？市场营销怎么样？

4. 我的服务器今天早上出了问题。

5. 众所周知，破坏比建造更容易。

6. 光化学反应将光转化为电脉冲。

7. 实验结果表明，该算法具有较强的抵抗正常攻击和几何攻击的能力。

8. 来自不同行业、不同领域的反馈结果不同。

9. 计划概念中的另一个基本考虑因素是功能。

10. 人口往往呈指数级增长。

Unit Three

【Ex.1】Content Questions.

1. Big Data technology stack can be broken down into two major components: the hardware component and the software component. The hardware component refers to the infrastructure. The software component can be further divided into data organization and management software, data analytics and discovery software, decision support and automation software. 大数据技术栈可以分为两个主要组件：硬件成分和软件成分。硬件成分指的是基础设施。软件成分可以进一步分为数据组织和管理软件、数据分析和发现软件、决策支持和自动化软件。

2. Infrastructure is the foundation of the Big Data technology stack. The main components of any data storage infrastructure (industry standard x86 servers and networking bandwidth of 10 Gbps) may be extended to a Big Data storage facility. 基础设施是大数据技术栈的基础。任何数据存储基础设施的主要成分（工业标准 x86 服务器和 10 Gbps 的网络带宽）都可以扩展为大型数据存储设施。

3. Two architectures — the extended Relational Database Management System (RDBMS) and the NoSQL Database Management System — have been developed to manage the different types of data. 为了管理不同类型的数据，开发了两种体系架构——扩展关系型数据库管理系统（RDBMS）和 NoSQL 数据库管理系统。

4. There are two decision support and automation software categories: transactional decision management software and project-based decision management software. The former is automated, embedded within applications, real-time and rules-based in nature. It enables the use of outputs to prescribe or enforce rules, methods and processes. Examples include fraud detection, securities trading, airline pricing optimization, product recommendation and network monitoring. Project-based decision management software is typically standalone, ad hoc and exploratory in nature. 有两类决策支持和自动化软件：交易决策管理软件和基于项目的决策管理软件。前者是自动化的，嵌入在应用程序中，本质上是实时且基于规则的。它允许使用输出来规定或执行规则、方法和进程。例如，欺诈检测、证券交易、航空公司定价优化、产品推荐和网络监控。基于项目的决策管理软件本质上通常是独立的、临时的和具有探索性的。

5. HDFS is a fault-tolerant storage system that can store huge amount of information, scales up incrementally and survives storage failure without losing data. Hadoop clusters are built with inexpensive computers. HDFS 是一个容错存储系统，它可以存储大量信息，逐步扩展并在不丢失数据的情况下避免存储故障。Hadoop 集群是用价格较低的计算机构建的。

6. Hadoop MapReduce's parallel processing capability has increased the speed of extraction and transformation of data. Hadoop MapReduce can be used as a data integration tool by reducing large amount of data to its representative form which can then be stored in the data warehouse. Hadoop MapReduce 的并行处理能力提高了数据的提取和转换速度。Hadoop MapReduce 可以作为数据集成工具，将大量的数据缩减为其代表形式，然后将其存储在数据仓库中。

7. Lack of industry standards is a major inhibitor to Hadoop adoption. Currently, a number of emerging Hadoop vendors are offering their customized versions of Hadoop. HDFS is not fully Portable Operating System Interface (POSIX) compliant, which means system administrators cannot interact with it the same way they would with a Linux or UNIX system. 缺乏行业标准是 Hadoop 应用的主要障碍。目前，许多新兴的 Hadoop 供应商都提供了他们定制的 Hadoop 版本。HDFS 不完全兼容可移植操作系统接口（POSIX），这意味着系统管理员不能像对待 Linux 或 UNIX 系统那样与 HDFS 进行交互。

【Ex. 2】句子翻译。

1. 使用此方法，每个开发人员都可以为此变量提供自己的物理路径定义。

2. 将所有数据转储到主计算机中。

3. 示例中已经实现了所有配置和代码。

4. 这是你可以选择的最简单的失重线程调度程序。

5. 这样的模型与敏捷的思维一致。

6. 这会导致各种可扩展性和维护问题。

7. 这使得存储节点可以在发现设备出现故障时复制数据。

8. 可扩展带宽提供了解决方案，同时提供了更有效的网络资源。

9. 冗余和可靠性给云带来了另一个优势。

10. 软件企业使产品本地化的最根本原因是增加总收入和净收入。

【Ex. 3】将下列词填入适当的位置（每词只用一次）。

（1）collected （2）special （3）source （4）individual
（5）completion （6）bottom （7）operations （8）machines
（9）node   （10）duplicate

Unit Four

【Ex. 1】Content Questions.

1. Text analytics is the process of deriving information from text sources. These text sources are forms of semi-structured data that include Web-based materials, blogs and social media postings (such as tweets). The technology within text analytics comes from fundamental fields including linguistics, statistics and Machine Learning. In general, modern text analytics uses statistical models, coupled with linguistic theories, to capture patterns in human languages so that machines can "understand" the meaning of texts and perform various text analytics tasks. These tasks can be as simple as entity extraction or more complex in the form of fact extraction and concept extraction. 文本分析是从文本源导出信息的过程。这些文本源为半结构化数据形式，包括网络资源、博客和社交媒体发布的帖子（如推文）。文本分析中的技术来自语言学、统计学和机器学习等基础领域。通常，现代文本分析使用统计模型和语言理论来捕捉人类语言的模式，以使机器"理解"文本的意义并执行各种文本分析任务。这些任务可以像实体提取一样简单，也可以以更复杂的事实提取和概念提取形式存在。

2. In-memory analytics is an analytics layer in which detailed data (up to TeraByte size) is loaded directly into the system memory from a variety of data sources, for fast query and performance calculation. In theory, this approach partly removes the need to build metadata in the

form of relational aggregates and precalculated cubes. 内存分析是一个分析层，在该层中，详细数据（大小高达太字节）从各种数据源直接加载到系统内存中，以实现快速查询和性能计算。从理论上讲，这种方法部分消除了以关系聚合和预计算数据库的形式构建元数据的需要。

3. Predictive analytics is a set of statistical and analytical techniques that are used to uncover relationships and patterns within a large volume of data that can be used to predict behaviours or events. Predictive analytics may mine information and patterns in structured and unstructured data sets as well as data streams to anticipate future outcomes. The real value of predictive analytics is to provide predictive support service that goes beyond traditional reactive break-and-fix assistance and toward a proactive support system by preventing service-impacting events from occurring. 预测分析是一组统计和分析技术，用于发现大量数据中的关系和模式，这些数据可用于预测行为或事件。预测分析可以挖掘结构化和非结构化数据集及数据流中的信息和模式，以预测未来的结果。预测分析的真正价值在于通过防止影响服务的事件的发生，来提供超越传统的被动中断和修复辅助的预测性支持服务，并向主动式支持系统发展。

4.（1）Pattern-based approach（基于模式的方法）

（2）Rule-based approach（基于规则的方法）

（3）Statistical process control-based approach（基于统计过程控制的方法）

【Ex. 2】补充空白部分的单词。

1. In general, modern text analytics uses statistical models, coupled with linguistic theories, to capture patterns in human languages so that machines can "understand" the meaning of texts and perform various text analytics tasks.

2. Text analytics will be an increasingly important tool for organizations as the attention shifts from structured data analysis to semi-structured data analysis.

3. The potential of text analytics in this application has spurred much research interest in the R&D community.

4. Applying text mining in the area of sentiment analysis helps organizations uncover sentiments to improve their Customer Relationship Management (CRM).

5. Adoption of text analytics is more than just deploying the technology.

6. On the traditional disk-based analytics platform, metadata has to be created before the actual analytics process takes place.

7. Developing a browser-based version of the business analytics application requires only a one-time development effort as the deployment can be made across devices.

8. Technical and security risk concerns will inhibit the uptake of mobile business analytics, especially in the case of mission-critical deployments.

Unit Five

【Ex. 1】Content Questions.

1. （1）Creating transparency.（创建透明度。）

   （2）Enabling experimentation.（进行实验。）

   （3）Segmenting populations.（细分人群。）

   （4）Replacing or supporting human decision making.（替代或支持人工决策。）

   （5）Innovating new business models, products and services.（创新商业模型、产品和服务。）

2. Big Data has a huge potential in the application of healthcare, particularly in areas where analysis of large data sets is a necessary precondition for creating value. Possible adoption of Big Data analytics could be done in a few specific areas. One of them is Comparative Effectiveness Research (CER). 大数据在医疗保健行业有巨大的应用潜力，特别是在那些将分析大型数据集作为创造价值的必要先决条件的领域。在几个特定领域可以采用大数据分析。其中之一是比较效益研究（CER）。

3. There are many ways which Big Data analytics can be applied in retail marketing. One of these applications is to enable cross-selling which uses all the data that can be known about a customer, including the customer's demographics, purchase history and preferences, to increase the average purchase amount. 大数据分析可以通过多种方式应用于零售营销。其中一个应用是启用交叉销售，交叉销售使用客户的所有已知数据（包括客户的人口资料、购买历史和偏好）来增加平均购买量。

4. （1）Personalized real-time information for travel options.（个性化的出行选项实时信息。）

   （2）Real-time driver assistance.（实时驾驶员辅助。）

   （3）Scheduling of mass transit system.（公共运输系统调度。）

   （4）Preventive maintenance.（预防性维护。）

   （5）Improved urban design.（改善城市设计。）

【Ex. 2】翻译句子。

1. 在这种情况下，你可以创建从存储库中的工作项检索数据的数据源。

2. 这是一个快速的方法，您可以根据自己的要求编辑和定制表单。

3. 成功利用此漏洞的攻击者可以作为登录用户运行任意代码。

4. 我们用这个晶体管放大电话信号。

5. 数据集还可能包含另一个存有订单信息的表。

6. 这在用例之间建立了一个工作流。

7. 因此，如果启用了安全性，则每个注册的基本节点可能配置了不同的用户注册表。

8. 较旧的机器需要加载一个软件修补程序来更正日期。

9. 通过提高自动化和效率可能会解雇行政人员。

10. 此时，你可以激活或停用任何你想要的其他插件。

【Ex. 3】将下列词填入适当的位置（每词只用一次）。

（1）organization　　（2）reputational　　（3）storage

（4）growing　　（5）relieves　　（6）mobile

（7）independently　　（8）requirements　　（9）exposure

（10）monitoring

【Ex. 4】补充空白部分的单词。

1. Big Data can be used to create value across sectors of the economy, bringing with a wave of innovation and productivity gains.

2. The discussion on the impact of Big Data focuses very much on the application of Big Data analytics rather than on the middleware or the infrastructure.

3. This data can be used to analyse variability in performance, identify root causes and discover needs or opportunities.

4. Though this functionality may be well-known in the field of marketing and risk management, its use in other sectors, particularly in the public sector, is not common and to a certain extent, may be considered revolutionary.

5. In some cases, decisions may not be completely automated but only augmented by the analysis of huge data sets using Big Data techniques rather than small data sets and samples that individual decision makers can handle and understand.

6. The common driver of Big Data analytics across verticals is to create meaningful insights that translate to new economic value.

7. Big Data has a huge potential in the application of healthcare, particularly in areas where analysis of large data sets is a necessary precondition for creating value.

8. The evidence is generated from research studies that compare drugs, medical devices, tests, surgeries, or ways to deliver healthcare.

Unit Six

【Ex. 1】Content Questions.

1. In a narrow sense, Cloud Computing is the delivery and use mode of IT infrastructure, which refers to obtaining the required resources in an on-demand and easy to expand way through the network; In a broad sense, Cloud Computing refers to the mode of service delivery and use, through the network to obtain the required services in an on-demand and easy to develop way. 从狭义上讲，云计算是 IT 基础设施的交付和使用模式，指通过网络以按需且易扩展的方式获得所需资源；从广义上讲，云计算指通过网络以按需且易扩展的方式获得所需服务的服务交付和使用模式。

2. Grid computing, utility computing, autonomic computing.

3. （1）Safe and reliable storage data.（存储数据安全可靠。）

（2）Low requirements of Cloud Computing for clients.（云计算对客户端的要求低。）

（3）Easy data sharing.（轻松共享数据。）

（4）It could be infinite.（无限的可能性。）

4. First of all, you only need to have a computer that can access the Internet. Secondly, there is a browser you like on your computer. Next, you just need to type the URL in the browser, and then you can enjoy the unlimited fun brought by Cloud Computing. 首先，你只需要有一台可以上网的计算机。其次，你的计算机上有你喜欢的浏览器。接下来，你只需要在浏览器中输入 URL，即可尽情享受云计算带给你的无限乐趣了。

【Ex. 2】补充空白部分的单词。

1. As you can see, your computer will be damaged or attacked, and some lawbreakers will steal your computer data by various means.

2. In a narrow sense, Cloud Computing is the delivery and use mode of IT infrastructure, which refers to obtaining the required resources in an on-demand and easy to expand way through the network.

3. This can be not only IT, software, Internet related services, but also applications in other fields.

4. Cloud Computing is a pay as you use model, which provides available and convenient on-demand network access to the configurable computing resource sharing pool (resources include networks, servers, storages, application softwares and services).

5. These resources can be provided quickly with little management or interaction with service providers.

6. In the past, people's traditional way of data storage is to store in the hard disk, many users suffer from virus, Trojan horse attacks, or due to some negligent operation resulting in data loss.

7. If the computer breaks down or even completely destroys, you can still recover the data through cloud on another computer.

8. At the other end of the cloud, a professional team will manage your information and save data for you.

Unit Seven

【Ex. 1】Content Questions.

1. As a new information technology, blockchain uses time stamp and digital password technology to record transaction records in the data block composed of time series, and uses consensus mechanism to store data in the distributed database, so as to generate the unique data record that is permanently saved and irreversibly tampered, and achieve the purpose of realizing credible transaction without relying on any central organization. 作为一种新兴的信息技术，区块链使用时间戳和数字密码技术，把交易记录记载在按时间序列组成的数据区块中，并使用共识机制把数据存储到分布式数据库内，从而生成了永久保存、不可逆向篡改的唯一数据记录，达到不依靠任何中心机构实现可信交易的目的。

2. The difference between blockchain ledger database and traditional database is just like the difference between postcard and ordinary mail in that era: the content of the postcard may be seen by many people in or around the post office, and the receiver and sender want to deny it, but they can't deny it. But ordinary letters are different. Outsiders don't know what they are about. At the same time, the data record of blockchain is unchangeable and permanent. 区块链分类账数据库与传统数据库的差别就像那个年代的明信片与平信的差别：明信片的内容可能会被邮局的人或周围的许多人看到，收信人和发信人想否认也否认不了；而平信则不同，外人并不知道信里写的内容是什么。同时，区块链的数据记录不可篡改且永久不变。

3. Through consensus mechanism, blockchain creatively solves the problem of information synchronization of all accounting nodes in the whole network. It can effectively get rid of the influence of some problem nodes and complete the correct accounting update. 区块链通过共识机制创造性地解决了全网各记账节点的信息同步问题，可以有效摆脱某些问题节点的影响，完成正确的记账更新。

4. The distributed storage of blockchain is to store all records in multiple accounting nodes of the whole network. The damage or loss of a single node will not affect other nodes, and the data error or tampering of a single node is more unlikely to have any destructive impact on the overall data. 而区块链的分布式存储就是把全部记录分布式保存在整个网络的多个记账节点上，单个节点的损坏或丢失并不会对其他节点造成影响，单个节点的数据错误或篡改更不可能对整体数据产生什么破坏性影响。

【Ex. 2】补充空白部分的单词。

1. A more vivid example is that in the era without network and telephone, people's remote information exchange is mainly through the post office.

2. The difference between blockchain ledger database and traditional database is just like the difference between postcard and ordinary mail in that era.

3. At the same time, the data record of blockchain is unchangeable and permanent.

4. This event typically reflects the huge risks of the centralized database.

5. The distributed storage of blockchain is to store all records in multiple accounting nodes of the whole network.

6. The damage or loss of a single node will not affect other nodes, and the data error or tampering of a single node is more unlikely to have any destructive impact on the overall data.

7. The data stored in different nodes are strictly protected by cryptography technology.

8. Through consensus mechanism, blockchain creatively solves the problem of information synchronization of all accounting nodes in the whole network.

Unit Eight

【Ex. 1】Content Questions.

1. Structured data can be stored in an ordinary database, and the corresponding processing will be more standard and simple. Unstructured data involves all kinds of information sources, and its proportion will rise sharply in the era of Big Data. 结构化数据可以存储在普通数据库中，且相应的处理会比较规范和简单。而非结构化数据涉及各种信息来源，在大数据时代，其比例将急剧上升。

2. Without the help of Artificial Intelligence, Big Data is difficult to do high-level processing and analysis. 没有人工智能的帮助，大数据很难完成高级处理与分析。

【Ex. 2】补充空白部分的单词。

1. Artificial Intelligence involves a wide range of fields, and has penetrated into people's lives.

2. Based on the accumulation and application of Big Data, people began to find some rules in it, which triggered the demand for analysis, and made the machine start to have ideas.

3. As hardware performance improves and computing resources become more and more powerful, Big Data breeds today's Artificial Intelligence.

4. AI is based on Big Data, but these resources are usually in the hands of giants.

5. Artificial Intelligence itself is the partner that Big Data must be combined for further development.

6. Big Data collection is not only standardized data, but also a large number of unstructured data and different dimensions of complex information.

7. Artificial Intelligence needs not only the analysis and interpretation of existing information, but also the establishment of active information acquisition and learning ability.

8. With the accumulation of more data and the support of better hardware level, Artificial Intelligence will develop stronger vitality.

Appendix D 参考资料

1. Ma Y, Wu H, Wang L, et al. Remote Sensing Big Data Computing: Challenges and Opportunities[J]. Future Generations Computer Systems, 2015, 51:47-60.

2. Spiess J, Joens Y T, Dragnea R, et al. Using Big Data to Improve Customer Experience and Business Performance[J]. Bell Labs Technical Journal, 2014, 18(4):3-17.

3. Zwick M. Big data in Official Statistics[J]. Bundesgesundheitsblatt - Gesundheitsforschung - Gesundheitsschutz, 2015, 58(8):838-843.

4. Mayer-Schönberger V. Big Data: A Revolution That Will Transform Our Lives[J]. Bundesgesundheitsblatt - Gesundheitsforschung - Gesundheitsschutz, 2015, 58(8):788-793.

5. Li M, Zhu L. Big Data Processing Technology Research and Application Prospects[C]. Harbin: International Conference on Instrumentation & Measurement, 2014, 7(2):269-273.

6. Benjelloun F Z, Lahcen A A, Belfkih S. An Overview of Big Data Opportunities, Applications and Tools[C]. Fez: Intelligent Systems & Computer Vision, 2015:1-6.

7. James M, Michael C, Brad B, et al. Big Data: The Next Frontier for Innovation, Competition and Productivity[N/OL]. Mckinsey Global Institute, 2011. http://www.mckinsey.com/business-functions/digital-mckinsey/our-insights/big-data-the-next-frontier-for-innovation.

8. 张强华, 刘俊辉, 郑聪玲, 司爱侠. 大数据专业英语教程[M]. 北京：清华大学出版社, 2019.

9. 吕云翔. 计算机专业英语[M]. 北京：电子工业出版社, 2019.

10. 明特·戴尔, 迦勒·斯托基. 未来之必然[M]. 北京：人民邮电出版社, 2019.

11. 朱丹, 王敏, 张琦, 陈宏. 大数据专业英语教程[M]. 北京：清华大学出版社, 2018.

12. 宋炜. 对话大数据[M]. 北京：人民邮电出版社, 2018.

13. 国泰安金融大数据研究中心. 大数据导论[M]. 北京：清华大学出版社, 2017.

14. 托马斯·埃尔, 瓦吉德·哈塔克, 保罗·布勒. 大数据导论[M]. 北京：机械工业出版社, 2017.

15. 武源文, 赵国栋, 刘文献. 区块链与大数据[M]. 北京：人民邮电出版社, 2017.

16. 周苏, 王文. 大数据导论[M]. 北京：清华大学出版社, 2016.

17. 赵伟. 大数据在中国[M]. 南京：江苏文艺出版社, 2014.

反侵权盗版声明

  电子工业出版社依法对本作品享有专有出版权。任何未经权利人书面许可，复制、销售或通过信息网络传播本作品的行为；歪曲、篡改、剽窃本作品的行为，均违反《中华人民共和国著作权法》，其行为人应承担相应的民事责任和行政责任，构成犯罪的，将被依法追究刑事责任。

  为了维护市场秩序，保护权利人的合法权益，我社将依法查处和打击侵权盗版的单位和个人。欢迎社会各界人士积极举报侵权盗版行为，本社将奖励举报有功人员，并保证举报人的信息不被泄露。

举报电话：（010）88254396；（010）88258888
传 真：（010）88254397
E-mail：dbqq@phei.com.cn
通信地址：北京市万寿路173信箱
    电子工业出版社总编办公室
邮 编：100036